My Search for Ramanujan
How I Learned to Count

Ken Ono • Amir D. Aczel

My Search for Ramanujan
How I Learned to Count

 Springer

Ken Ono
Department of Mathematics
and Computer Science
Emory University
Atlanta, GA, USA

Amir D. Aczel
Center for Philosophy & History of Science
Boston University
Boston, MA, USA

ISBN 978-3-319-25566-8   ISBN 978-3-319-25568-2   (eBook)
DOI 10.1007/978-3-319-25568-2

Library of Congress Control Number: 2015959483

Springer Cham Heidelberg New York Dordrecht London
© Springer International Publishing Switzerland 2016

Printed on acid-free paper

Springer International Publishing AG Switzerland is part of Springer Science+Business Media (www.springer.com)

To the memory of Basil Gordon and Paul Sally and to Andrew Granville

PREFACE

Although this book is a first-person narrative in my voice, it represents collaborative work between me and my friend and coauthor, Amir D. Aczel. This book was his idea, and without him it would never have been written. I am deeply saddened by the fact that Amir passed away unexpectedly before work on the book was complete. Amir was a passionate man who took great pride in spreading scientific ideas through his writing. I will miss him. He still had so much to give to the world.

Ever since I was a teenager growing up in a suburb of Baltimore, I have been enchanted by the story of Ramanujan. I first learned about him when my father, a prominent mathematician, received a letter from Ramanujan's widow, sixty years after her husband's death. Srinivasa Ramanujan Iyengar was an enigmatic Indian mathematician who, without formal education in advanced mathematics, was able to derive thousands of unproved, yet valid, mathematical formulas and identities. As a mathematician, I have spent my career extending and proving some of Ramanujan's results, largely in an attempt to understand the soul and spirit of this genius who died at the young age of thirty-two. Along the way, I found myself and came to terms with my past.

This book is the result of my strong need to tell the world about how the story of this man and his mathematics helped transform me. I was once an emotionally frail, dispirited sixteen-year-old high-school dropout on the run from myself. Today, I am content. I have a loving family, I am a successful mathematician, and I have a rich spiritual life. As the character Jerry puts it in Edward Albee's play *The Zoo Story*, "Sometimes a person has to go a very long distance out of his way to come back a short distance correctly."

Acknowledgments

Amir and I could not have written this book without the help and support of many people. First of all, we thank our families: our wives, Debra Aczel and Erika Ono, and my parents, Sachiko and Takashi Ono, for their tough love and unwavering support. I thank my colleagues Krishnaswami Alladi, George

Andrews, Dick Askey, and Bruce Berndt for their shared enthusiasm for this story. We express our deepest gratitude to SASTRA University, in India, for converting Ramanujan's childhood home in Kumbakonam into a museum and for establishing the SASTRA Ramanujan Prize. I am indebted to Krishnaswami Alladi for playing a central role in honoring the memory of Ramanujan and for hosting me on my many trips to India. I thank Emory University and the Asa Griggs Candler Fund for their financial support, and we thank Matthew Brown, director of the film *The Man Who Knew Infinity*, and Pressman Films for their cooperation. The film is based on the superb book *The Man Who Knew Infinity: A Life of the Genius Ramanujan*, by Robert Kanigel. We are grateful to Robert for writing this exceptional biography twenty-five years ago.

We are indebted to Henna Cho, Carol Clark, Melissa Mouly Di Teresa, Danny Gulden, Robert Schneider, Marc Strauss, and Sarah Trebat-Leder, who provided many helpful comments on earlier drafts of this book. Without their help, the book would have fallen far short of its intended goals. We thank our editor, Marc Strauss, for encouraging us to tell this story. And we thank our copyeditor, David Kramer, for beautifying and polishing our manuscript. He improved our book in uncountably many ways. Finally, we thank our literary agent, Albert Zuckerman, for helping us make this book a reality.

I am one of the luckiest mathematicians in the world. I have been guided by three amazing men, without whose friendship and guidance I would certainly have had nothing to write about. We dedicate this book to them, my mentors: Paul Sally, Basil Gordon, and Andrew Granville. Sally rescued me when I was an unmotivated undergraduate at the University of Chicago. Gordon taught me how to do mathematics for its own sake. Granville taught me how to become a professional mathematician. These men reformed, transformed, inspired, and coached me, and they made me what I am today, an active, spiritually aware mathematician with a story to tell.

Atlanta, GA   Ken Ono
Boston, MA   Amir D. Aczel
December 2015

PROLOGUE: MY HAPPY PLACE

I cannot sleep. I have a lot on my mind. It is 5:30 a.m. on May 28, 2015, and I am sitting on the lanai of an oceanview room at the Makena Beach and Golf Resort on the island of Maui. The doves are cooing. The gentle ocean breeze and the soothing sounds of waves crashing on the white sandy beach below define this heavenly moment. I am waiting for the first hint of the sun's rays in what I expect will be an absolutely glorious sunrise, with the Haleakala volcano as a backdrop.

I have never been happier.

Last night, I enjoyed a lovely evening with my wife, Erika, complete with a delicious sunset dinner at a fancy Italian restaurant on the beach in Kihei. This week we are celebrating our twenty-fifth wedding anniversary, doing many of the things we love most: mountain biking in the Makawao Forest Preserve, scuba diving in search of manta rays and green sea turtles, hiking the lava fields that formed this gorgeous island, surfing the waves at Kalama Beach, among other activities that define our active lives. Our teenage children, Aspen, who is an undergraduate at Emory University, and Sage, a rising junior at Centennial High School, in Roswell, Georgia, are home alone, enjoying ten days of freedom from their parents. They are great kids; we couldn't be more proud of the young adults that they have become.

Although the last twenty-five years seem to have passed by in a blur, Erika and I are blessed in that life has not passed us by. We have spent much of this week reminiscing about the path we have taken—our college years in Chicago, our years in Los Angeles, where I earned my doctorate, my postdoctoral tour of America, the birth of our kids, and so on. We have been leading rich lives. We have traveled the world, and we have many good friends. We are at peace with who we are as a couple, as parents, and as individuals. I am a successful mathematician, a professor at Emory University, and I am considered a leader in my field. I am a well-known mentor of young mathematicians. To borrow an overused phrase, "I'm living the dream."

How have I been lucky enough to get to where I am today?

With Erika on Maui (May 27, 2015)

That is the question that keeps me awake. It is a question of grace and gratitude, a question that I ask every day in wonder, almost out of fear that I will be awakened from this lovely dream to discover that none of my life has actually happened. It is a question that will haunt me for the rest of my life. Erika knows this about me. But even she will be surprised when she reads some of the details of the story I am about to tell, events that I haven't shared with anyone before.

It wasn't that long ago that my waking thoughts, the parental voices in my head, were about inadequacy and fear of failure. There was a time when I couldn't even imagine wanting to live long enough to witness my thirty-second birthday. In a moment of weakness and despair twenty-three years ago, when I was twenty-four, I came within seconds of taking my own life. In a torrential rainstorm, I veered over the double yellow line of a Montana highway, near a place called Ronan, with the intention of driving headlong into an oncoming logging truck. I swerved back into my lane at the sound of the trucker's frantic horn, and I pulled over and came to a stop on that lonely strip of asphalt in the middle of a vast wilderness. I sat there alone, with the engine running, protected from the driving rain, for an eternity, trying to figure out what I was going to do with the life I had just failed to destroy. The harrowing voices, the product of my confusing and frustrating childhood as one of three sons of tough-loving, hard-driving Japanese-American parents, had nearly dragged me to my death.

Most of my current acquaintances know nothing of my past. I have done my best to erase my childhood by pretending to myself that my life began when I was sixteen years old. I ran away from my former life, and I have successfully managed to live my adult life in self-imposed amnesia. I almost never talk about my life before college; the memories are too painful. Instead of coming to terms with the past, I dumped it into a black hole, a massive part of my history from which memories could not escape.

I was so tortured by those voices that I dropped out of high school in the naive hope that I could escape my feelings of inadequacy. I have spoken to not more than one or two of my former classmates and friends since that day over thirty years ago. I didn't return phone calls from former friends who were wondering what had become of me. I made a clean break. I simply abandoned my former self. Although I didn't understand it at the time, I now feel that I was committing a metaphorical ritual suicide, what the Japanese call *seppuku*.

I was raised by "tiger parents," a term applied to Asian-American parents who raise their children in an overly strict way (by Western standards) with the goal of fostering an academically competitive spirit. This form of upbringing is intended to direct a child toward success, and it has one goal—to raise extraordinarily successful children, conquerors of their fields. Tiger parents set high academic standards, and they severely restrict nonacademic activities with this long-term goal in mind. That is how my parents showed their love for me, by demanding nothing less than the best. But the cost was enormous. It almost killed me.

My professional success confirms, and perhaps even justifies, the merits of this philosophy. There is certainly much truth to the undeniable fact that my parents fostered qualities in me that have been essential for my success. I am ambitious and competitive. I am restless, anxious to take on the next challenge. I thank my parents, whom I love deeply, for instilling these qualities in me.

This style of parenting, however, especially when carried to the degree that I experienced, has the potential of crushing a child's spirit. Children can grow up emotionally unfulfilled and starving for recognition. Those voices in my head that nearly drove me mad, my despair and feelings of worthlessness, all attest to that fact.

What vanquished those voices and gave me back my soul? The answer involves a mysterious genius from India, a man whose story inspired me at my lowest points, and whose ideas have powered my career.

Let me tell it to you as I remember it.

Makena Beach, Maui, HI
May 2015

CONTENTS

CONTENTS

CONTENTS

CONTENTS

Part I
My Life Before Ramanujan

Chapter 1

TIGER BOY

Lutherville, Maryland (1976–1984)

I'm sitting on the couch watching *Gilligan's Island*, every second-grader's favorite sitcom. It's the episode where the headhunters from a neighboring island attack the motley crew of castaways. As usual, the klutzy skinny first mate Gilligan accidentally saves the day, in this episode by scaring off the headhunters.

I really should be doing the geometry problems that my parents assigned me, but they aren't home, and I love *Gilligan's Island*. If my parents find out, I'll be in super big trouble. But I'm prepared. I have a washcloth and a small pink plastic basin filled with ice water. The TV is on low enough that I will be able to hear my parents pull into the driveway, which will give me just enough time to turn off the TV and cool the back of the set with the ice-cold washcloth.

My second-grade portrait

© Springer International Publishing Switzerland 2016
K. Ono, A.D. Aczel, *My Search for Ramanujan*, DOI 10.1007/978-3-319-25568-2_1

Does this seem over the top?

In her book *Battle Hymn of the Tiger Mother*, Amy Chua, a Yale law professor, writes about what she calls tiger parenting, a traditional strict form of child-rearing popular in Asia and among Asian-American families. This "tiger mom" ideology accurately describes the approach of my Japanese immigrant parents, Sachiko and Takashi Ono. Like other tiger parents, they believed that their children could be "the best" students and that academic achievement is a reflection of successful parenting. Indeed, if their children are not at the top of their class, then the parents aren't doing their job. My parents went a step further. If I wasn't the best student, then I would bring shame on my family. It was understood that it was my duty to be "the best."

My parents in 1999 (photo by
Olan Mills)

I emerged from my early childhood with the voices of my parents in my head that continually rebuked me for my inadequacy and my inability to live up to their unrealistic expectations:

> *Ken-chan, your parents are disappointed in you. You are embarrassment.*
> *Look at that professor's children. Unlike you, they study all of time, and they*
> *what you should be. You sloppy. You spoiled. Your mother sacrificed her life*
> *for you, so you do your part. What wrong with you? You want play all of time?*

Those voices told me that my parents would love me only if I was both a star student and a brilliant musician. Those voices told me that it was wrong to relax and have fun and hang out with friends. When I did those things, those voices made certain that I suffered tremendous pangs of guilt.

I now understand that many children today hear similar voices. Tragically, some of these children will succumb to those voices and take their own lives. Moreover, those suicides often occur in clusters, a phenomenon that has recently become a source of concern in communities like Palo Alto, where elevated academic expectations are rampant and such parenting is common.

Those voices are symptoms of an anxiety disorder that has been the focus of considerable recent study by clinical psychologists. Their research suggests that children of tiger parents are often burdened with anxieties that last a lifetime. The research also offers a possible biological explanation for this phenomenon.

When we realize that we have made a mistake, a predictable electroneurological process called *error-related negativity* (ERN) is triggered in the medial prefrontal cortex of our brains. It acts as a reset button for the brain. It is now believed that the strength of ERN is negatively impacted by prolonged exposure to harsh criticism.

I didn't need this research to understand the validity of its conclusions. I have firsthand knowledge. I became desperate for the love and approval of my parents, and when I failed and failed again to obtain it, my life began to unravel.

For you to understand how all this came to be, I will have to explain my family. My parents raised my two brothers and me under the assumption that we were somehow genetically predestined, with each son to follow a well-defined path that my parents determined in response to the talents and strengths we exhibited in our early years. Our job was simple—stay on track and succeed in the lives that our parents had prescribed for us. I felt that I never had a choice.

My oldest brother, Momoro, was gifted in music. He was a child prodigy. You know the kind—the cute Asian-American third-grader with a bowl haircut, dressed in a tuxedo, dazzling television audiences with a precocious rendition of a Tchaikovsky piano concerto. He was going to be a concert pianist who performed at Carnegie Hall.

Santa, the middle son, had a different path. He was often described as the black sheep of the family, which is ironic, because he is the one who will go on to be the most successful son. My parents felt that he was unlikely to amount to much of anything, so he was expected to be an ordinary company man, whatever that meant. As a second-grader, I understood only that it referred to something that my parents viewed as significantly below concert pianist and university professor.

I was being groomed to be a mathematician in the image of my father. But I was also expected to be an outstanding musician. I was only in second grade, but I already had the next twenty-five years mapped out for me. I was to attend one of the best Ivy League universities, earn a PhD in mathematics, and then secure a professorship at a top university.

My parents showed their love for us, which I didn't understand at the time, by defining our long-term professional goals and offering opportunities that powered us toward them. Their entire focus was on those goals, in the belief that we would reach our happy places by achieving them.

The rules that they made were simple to follow, reducing each of our lives to an individual formula. To achieve the goals that our parents had set for us was going to be easy—as easy as basic algebra: just as plugging $x = 2$ into the formula $y = 3x + 1$ gives the value $y = 7$, I was given a simple formula for becoming a mathematician—I had to be a straight-A student who earned top scores on all my tests. And for good measure, I was expected to become an accomplished violinist.

Everything outside of the formula was considered extraneous, and if we ever strayed in the slightest from our formulas, we were subjected to a litany of rebukes and threats that discouraged and humiliated us.

Here is a vivid example of what I am talking about. I was in third grade. Each year in elementary school, we took the Iowa Tests of Basic Skills, standardized tests that were administered as a tool for improving instruction. These tests evaluated our skills in grammar, reading comprehension, and mathematics. When my parents received my scores, they were shocked at my poor performance: ninety-eighth percentile in math and ninety-seventh percentile in reading comprehension. They summoned me to the kitchen and sat me down at the head of the table. Pacing behind me, they rebuked me for my embarrassingly inadequate performance:

Ken-chan, one in fifty people did better in math. Even more better in English. That's fifty thousand kids in country. You aren't even among best fifty thousand. Harvard only accepts few thousand. If you don't fix, you might end up at University of Maryland, or, God forbid, Towson State. You must get ninety-ninth percentiles. That professor's children all got these scores, and look how successful they are. We sacrificing everything for you three boys, and this how you thank us? If you won't do better, then get out of house!

Then they left the room, while I sat alone at the table crying, with only the ticking of the clock and the rumble of the fridge to keep me company. Half an hour later I'd hear the shuffling of slippers, which signaled their return, and they repeated the diatribe again. I wanted to sink through the floor, or at least to escape to the privacy of my room.

Between me and my two brothers, incidents like this were common in my early childhood. One of us would be scolded while the other two would cower in their bedrooms, thankful that it wasn't their turn. Truth be told, I was on the receiving end of these rants much less often than my two older brothers.

My parents forbade just about everything that was not directly tied to our formulaic lives. There was no room for *Gilligan's Island*. Santa was forbidden from attending his high-school prom. When he sneaked out to attend despite their prohibition, my parents tracked him down and brought him home. Imagine the humiliation of being pulled from the prom in front of your classmates. My parents' view was that this time-honored rite of passage was not for serious students, not part of the formula. Perhaps they also wanted to protect him from the bad influences they imagined he would be exposed to.

Here is another example. Let me explain why I can't watch the *Star Wars* movies. When the first *Star Wars* film came out in 1977, I begged my parents to let me go with my friends to the movie theater to see it. Many of my friends saw the film several times, and so I had plenty of opportunities. After incessant pleading, my parents finally relented, but under one condition—that I write a three-page essay explaining the importance of the film and the deeper meaning of the story. I was a nine-year-old who simply wanted to delight in the droids and bizarre space creatures that everyone was talking about. Instead, the film became an academic exercise complete with edits and revisions. To this day, I can't watch any of the sequels. The mere thought of *Star Wars* stirs up painful memories.

But when it came to academics and music, my parents offered me the best. They bought me computers: I was one of the first kids at school to get an Apple II. They enrolled me for violin lessons at the Peabody Institute. Peabody is an internationally renowned conservatory that has trained famous musicians like singer Tori Amos and pianist André Watts. My parents engaged Yong Ku Ahn, a distinguished Peabody Institute professor, to teach me to play the violin. And the violin they bought for me was the product of a nineteenth-century Italian craftsman. I didn't understand until much later that beginning violinists don't usually take their first lessons from a well-known virtuoso, and they don't typically play an instrument whose value is roughly that of a brand-new car.

As part of my mathematical training, my father had me write computer programs to collect data for his mathematics graduate students. I actually enjoyed doing this, and it was the source of some of the very few memories I have of the parental attention and approval that I so badly craved. Even though I didn't know anything about class numbers of quadratic forms, I was able to write computer programs that computed tables of them. I was delighted that I, a kid in elementary school, could help produce results that were published in a PhD thesis.

But these opportunities always came with a price, and when it came to music, the price was more than I was able or willing to pay. I took lessons from Professor Ahn for almost ten years. Each week, both my parents drove me to his home, and then they sat in on my lesson, taking note of my progress. The pressure was unendurable. Instead of concentrating on my playing, I constantly had my eyes on my parents, looking for evidence of their approval or chagrin. They never told me that I had done well; I believe that the thought of giving praise never crossed their minds. And on the long drive home, they would rehash the lesson, reminding me of the mistakes I had made and emphasizing the improvements that were expected of me for the next lesson.

Despite the fact that I became an accomplished violinist, making second chair in the first violin section of the Peabody Preparatory Orchestra, there is very little that I can say that might put a positive spin on my musical career. I hated the violin.

One day, when I was in tenth grade, I simply quit. I have picked up a violin only once since then, and that was five years later, when Erika, who would later become my wife, brought me to her home for the first time to meet her parents. In my desire to impress them, I unthinkingly broke my black-hole rule: I mentioned that I had been an accomplished violinist. To my surprise, they actually wanted to hear me play. And to my chagrin, there was a violin—Erika's sister's—in the house. They no doubt thought that I would be delighted to play for them, and they had every reason to expect to hear an "accomplished violinist" produce lovely music. I tried to play, but I couldn't. Instead, I sat alone in Erika's sister's bedroom, holding her violin while I tore myself apart inside. How could I possibly explain? How could I make them understand? They had no idea of what they were asking me to do.

Still in the bedroom, I finally worked up a bit of courage and began softly playing Bach's *Partita Number 3* from memory. The sight of my reflection in the mirrored closet doors aroused painful long-dormant memories, and it was more than I could bear. After a few minutes, I stopped playing, and I told Erika and her parents that I just couldn't perform for them. At least, I thought, they had overheard enough to know that I had indeed been a competent violinist. And I was grateful that they had apparently recognized that I was in the grip of an inner struggle, and I was relieved when they let the incident pass without pressing me for details. When this book is published and Erika's parents read this paragraph, they will finally understand something of the inner torment that I was suffering on that day almost thirty years ago.

Quitting the violin was my first successful act of rebellion against my parents. They berated me for the decision, and it took several weeks before they retreated in the face of my implacable conviction. We had heated arguments, and I heard many reasons why I had to resume my lessons:

If you quit, then what will everyone think? You disappointment for family. You let Professor Ahn down. He will never forgive you. He believe that you talented enough for Juilliard School. How can you quit after all that he done? He spent hundreds of hours when he could have taught someone else who grateful. If you quit he will cry. You understand how lucky you are? How can you quit something you good at? If I had your talent I would practice ten times as much as you. How could you be so thankless? We struggled to pay for expensive violin and lessons. Your mother took part-time job as seamstress doing alterations for local dry cleaner so that you can have violin and lessons. She could have go to beauty salon, but more important to make you something good for your life.

Although I had hell to pay for quitting the violin, I had stood up for myself, and I was proud for having had the courage and strength to do so. I had no idea that I would soon draw on that strength to rebel again by running away from my life.

I know now that my parents loved me. What I didn't understand at the time is that they showed their love in ways that I didn't, couldn't, recognize or appreciate. It was love, but it wasn't enough. I was completely unaware of their personal history as immigrants to America who had come of age in Japan during World War II and the postwar reconstruction. My parents rarely spoke of their childhood, and we had almost no contact with our relatives in Japan. I had no way of comprehending the demons and challenges that they faced as they did their best to raise three boys, isolated in a culture that they knew nothing about. I understood only that life—with my parents at the center—was not treating me fairly, and for that my heart was full of bitterness.

In my memory I was never hugged by my parents, and they never told me that they loved me during my formative years. I understand now my parents' rationale for how they raised me and my brothers. Their parenting focused exclusively on long-term professional goals, and they believed that offering praise and love for smaller accomplishments would diminish our chances of achieving the more worthy ones. They wanted their boys to be hungry for success, and so they starved us of praise. They aimed to foster our competitive spirit in this way, a common practice among tiger parents.

Of course, I didn't understand this as a teenager. I couldn't get beyond my pressure cooker of a life, and as one of the few Asian-American kids at school, I felt alone, without anyone to identify with. I was being tossed around between two divergent cultures. At school, I was a star student, while at home, I was a disappointment to my parents and a failure for falling short of their impossible expectations. And like any child, I wished to please my parents, but nothing I did was good enough for them. No matter how hard I tried, my accomplishments were ignored or belittled. They saw no point in acknowledging such insignificant achievements as straight A's on a report card or a medal from the local science fair.

My parents did not approve of my friends. Without a cohort of Japanese-American classmates at school who also happened to be the offspring of university professors, there was no way that I could have more than one or two friends that they were willing to embrace. In their view, it wasn't even important to have friends. It was enough that we had a dog, a dachshund, named Igor Stravinsky in honor of my father's favorite Russian composer.

My parents never said so, but I think that through the end of elementary school, I did my job well. I was one of the best students. But it wasn't easy; plugging into my formula was an emotionally exhausting occupation.

Beginning in seventh grade, my life became more of a struggle. My math teacher, Mrs. Sprankle, a wonderful woman in her fifties, encouraged her pupils' parents to be involved with her class. To promote this, she required us to show our parents our graded tests and return them to her with a parental signature. Although I was the top student in class, I couldn't bear to show my parents any test that had less than a perfect score. I knew what they would say. After all, my job was not to be the best student in class, it was to be a perfect student, and for the son of a famous mathematician, nothing less than perfection would do. We are not talking about higher mathematics here. This was first-year algebra, a subject that my parents didn't even view as mathematics. It was barely a step above glorified counting. As I write this, I hear echoes of their voices from the past. In a slow drawl with a Japanese accent they are saying, "Ken-chan, you got ninety-five percent. Not good. Why you get easy problem wrong? You must do better."

Despite the fact that I earned an A on every algebra test that year, I never showed a single one of them to my parents. Test after test, I locked myself in the bathroom and stared at the test paper trying to summon the courage to ask my mother for her signature. I was good at doing what I was told, but I could not do what Mrs. Sprankle had asked of me. Yet the test had to be signed, and so I deceived Mrs. Sprankle by forging my mother's signature again and again. The voices in my head began around that time.

Mrs. Sprankle was very fond of me. I was her best student. I spent hours visiting with her after school. I cleaned her chalkboard. I tutored other kids in class. She offered me encouragement and love. I didn't understand it at the time, but I realize now that she was nurturing me in ways that she must have seen I desperately needed.

I was so ashamed to have to deceive her, for she was one of the few adults in my life at the time who offered me encouragement and recognition. I couldn't understand how I could be the best math student at school, earning the praise and attention of my teachers, but then be not quite good enough at home. Not quite good enough? At home, I was an abject failure!

By the end of tenth grade, I couldn't take it anymore.

I awoke each day with very painful thoughts: I will never be good enough. I'm an impostor. My parents will never love me because I will never live up to their expectations. No matter how well I do, I ought to have done better. It seems that there is nothing I can do to earn their approval.

And so I dropped out.

Chapter 2

⁓

MY PARENTS' GENERATION

*I*f you are to have a sympathetic understanding of my history, you will have to understand my roots, the story of how my parents left their families and everything they knew behind when they came to America for what was supposed to be a short visit. That short visit became a lifetime, and my parents became *issei*, "first-generation" Japanese immigrants to the United States of America. And as the American-born sons of my parents, my brothers and I are *nisei*, children of the second generation.

Takasan, Tokyo (1943)

The slight fourteen-year-old-boy with angular facial features is riding his forty-pound one-speed Japanese bicycle, fenders creaking, through the crowded narrow streets of his Tokyo neighborhood. Dressed in his military-style uniform, he is riding to school on his daily route, passing bustling markets selling rice, tea, vinegar, and other goods at the start of what had begun as an ordinary school day. The streets are teeming with bicycles and pedestrians, all traveling in a frenetic but somehow orderly fashion. Many of the men are dressed in bland western business attire as they head for work, but the women are radiant in their colorful flowery kimonos.

The two-wheeler gives the boy a sense of freedom. He rides his bike ten miles every school day, safely weaving through the commotion that is Tokyo. Those miles of solitary travel in Tokyo's tumult offer him a private world, an island of solitude where he can escape boredom as the time flies beneath his wheels.

But this morning turns out to be no ordinary day. Suddenly, his rear wheel kicks up a large stone and flings the projectile in a fateful arc straight through the large plate glass window of a small tailor shop, smashing it to bits. He halts his bike.

© Springer International Publishing Switzerland 2016
K. Ono, A.D. Aczel, *My Search for Ramanujan*, DOI 10.1007/978-3-319-25568-2_2

The thunderous sound of crashing shards and the crowd of onlookers that gathers frighten him. In a panic, and afraid of facing the shopkeeper's wrath, the boy mounts his bicycle and disappears down the crowded street.

After school, he chooses a different route home; he is afraid to face the consequences. Later that night, overwhelmed with shame and guilt, he vows to visit the shop the next day to pay for the repairs. The next morning, after some further soul-searching, he musters his courage and makes his way to the shop.

To his horror, he arrives to find a giant crater where the shop had been. The previous night, as was common at that time, a fleet of American bombers had conducted a long-range firebombing raid. The shop has disappeared, as has almost the entire block. There is no shop-owner to repay; he is one of many innocent casualties of war whose lives have been extinguished for reasons the boy doesn't understand.

That boy is Takashi Ono, and he will become my father, whom we call *Takasan* in reverence.

This story is one of the few that my father tells about his childhood. My parents were born and raised in Japan. My father was born in 1928, and my mother in 1936. They came of age during World War II and the subsequent difficult postwar reconstruction.

Takasan, ca. 1935

My father has a small number of stories about his childhood that he likes to tell and retell. Some involve his athletic prowess; it seems that he was a fast runner in high school. However, the crucial ones, the stories that must be understood to penetrate our family, all involve war. They are stories about fire and smoke, weeks of hunger, and the stench of death. They are the stuff of nightmares, but for my parents, they are more than bad dreams. Those nightmares are their history, their daily reality when they were growing up in Yokohama and Tokyo at a time when surviving to the next day was the only goal.

It was difficult for the Japanese to grasp what was happening. Their country was supposed to be everywhere triumphant. In the eighty years preceding the war, Japan had transformed itself from a feudal state to a world power thanks to a major investment in industrialization and modernization.

That development was accompanied by a rise in militarism, which led to Japan's quest to conquer Asia, beginning with victorious wars against China and Russia. During those conflicts, the Japanese military committed unspeakable atrocities against the civilian populations. The Nanjing Massacre, also known as the Rape of Nanjing, is perhaps the most infamous war crime committed by the Japanese military in the years immediately preceding World War II. Nanjing, the capital of the Republic of China, fell to the Japanese army in December 1937. For six weeks, Japanese soldiers raped, looted, and murdered Chinese citizens with impunity. The murders were justified by the pretext that large numbers of Chinese soldiers were disguised as civilians. It is estimated that 200,000 civilians and prisoners of war perished.

Such atrocities contributed to the worldwide opinion, which persists in the minds of some today, that the Japanese are a barbaric race. But the Japanese were taught otherwise, that they were superior to all other races.

In the nineteenth century, Japan had a spiritual government, the "State Shinto," which promoted the unity of government with certain aspects of Shintoism. The Japanese emperor was considered divine, and government leaders often performed religious ceremonies in which they communicated with the Sun Goddess. Japanese citizens were taught that self-sacrifice was the greatest virtue, and sacrificing oneself in service to the nation was regarded as the greatest expression of patriotism.

In the early twentieth century, as an extension of some of the principles underlying its isolationist policies of the mid-seventeenth through mid-nineteenth centuries, the imperial government reinforced the notion that the Japanese were an exceptional race. Government scientists even offered proof of this superiority by noting that the Japanese had a higher forehead-to-nose ratio, less body odor, and less body hair than other peoples. In contrast, the Americans were little more than *yaju,* wild beasts.

The Japanese government maintained this propaganda throughout World War II, even as firebombs devastated the neighborhoods of Tokyo. That was the Japan of my parents' youth, a tumultuous time when everything they had been taught about their country crumbled before their eyes.

For a people that had been taught that their emperor was a god and that their race was superior to all others, this change in fortune was unfathomable. Finally, they were confronted with the undeniable reality and humiliation of resounding defeat and unconditional surrender. What happens to a nation's psyche when its divine leader and superior race of warriors are defeated by a nation of wild beasts? How does a nation recover?

A nation recovers ultimately from the recovery of each individual, and for my father, the path to wholeness was through mathematics. Perhaps mathematics was his escape from a brutal and inhuman existence. He had always been good at math, and as a young man, he joined with other young Japanese men to create a community of self-taught mathematicians. Their goal was to escape their dreary lives by making a mark in the world of mathematics. They had to learn from one another: their universities had been decimated by war, and the few professors remaining were unaware of the current research that was being produced in Europe and the United States.

My father got his big break thanks to the generosity of André Weil, who had emigrated from France to the United States in 1941 and was now a member of the Institute for Advanced Study, in Princeton, New Jersey.

Weil was one of the most influential mathematicians of the twentieth century. He was a founding member and the self-styled leader of the Bourbaki group, a collective of mostly French mathematicians who aimed to reformulate mathematics based on extremely abstract but self-contained formal foundations. They wrote a series of books, published under the pseudonym Nicolas Bourbaki, that codified several branches of modern mathematics.

Weil met my father in 1955 during a trip to Japan to attend the Tokyo–Nikko conference on algebraic number theory, at which time my father was a graduate student at Nagoya University. The conference was an important opportunity for Japan to reengage with the world in the name of science. In his opening address, Zyoiti Suetuna, the chairman of the symposium, proclaimed,

We can say that this kind of an international meeting, which aims at the cultural development of mankind regardless of all the political, social,

and racial differences, will exert not a small influence on advancement of
international cooperation and peace in the world.

The conference was a tremendous success in this regard. In addition, the conference was the unexpected site of one of the most famous events in twentieth-century mathematics, though it would be recognized as such only many years later.

At the time of the symposium, Weil was a professor at the University of Chicago. Weil and Jean-Pierre Serre, a fellow Bourbaki member and future professional acquaintance of mine who had been awarded the Fields Medal the previous year, were two of the distinguished speakers at the conference. The Fields Medal, which is awarded at the quadrennial International Congress of Mathematicians, is regarded as the "Nobel Prize" of mathematics. The self-taught young Japanese mathematicians were thrilled to have the opportunity to mingle and chat with some of the world's best mathematicians.

For many of those young Japanese mathematicians, including my father, that meeting changed the course of their lives. These men had been toiling away for years on their research in solitude and isolation, first during World War II and then during the postwar reconstruction. They were poor, and they were hungry, and not only for knowledge, for they had little to eat, often getting by on nothing more than an occasional bowl of rice. Despite their circumstances, they set their sights high, and they aimed to make a mark in the world of mathematics. The conference was their opportunity to share the fruits of their labor with important mathematicians with access to the entire world of mathematics outside of Japan. These young men hoped finally to be rewarded with international recognition.

That dream became a reality for a surprisingly large number of them. Impressed by their work ethic and their accomplishments, Weil used his influence to arrange scientific opportunities for some at places like the Institute for Advanced Study. Thanks in part to his generosity, the best among them eventually secured professorships at top universities in the United States, the nation that Japan had ruthlessly attacked on December 7, 1941, at Pearl Harbor. That cohort included Jun-Ichi Igusa (Johns Hopkins), Shoshichi Kobayashi (Berkeley), Michio Kuga (SUNY Stony Brook), Ichiro Satake (Berkeley), Goro Shimura (Princeton), and Tsuneo Tamagawa (Yale). My father, who gave a lecture on orthogonal groups at the meeting, eventually landed a prestigious position at the Johns Hopkins University.

Part of the group photo at the 1955 Tokyo–Nikko conference (*Top row*: second from left, Jean-Pierre Serre; third from left, Yutaka Taniyama. *Third row*: second from left, Takashi Ono. *Second row*: on left, Goro Shimura. *Bottom row*: third from right, André Weil)

Anti-Japanese sentiment in the 1940s

Coming to America

Every family has its story, and my parents' reads like a fairytale. It is the story of a young starving Japanese couple who sought a better way of life than could be had in postwar Japan. My father, an aspiring mathematician, impressed the distinguished mathematician André Weil, who then provided the opportunity of a lifetime, the chance to study and work among the best in America, arranging a research position at the Institute for Advanced Study, "Einstein's institute in the woods."

The Institute for Advanced Study is one of the world's leading centers for theoretical research. It promotes curiosity-driven research in the humanities and the sciences. The Institute was founded in 1930 by the American educator Abraham Flexner with funding by the philanthropist businessman Louis Bamberger and his sister, Caroline Bamberger Fuld. Its picturesque campus occupies eight hundred acres of former farmland outside Princeton, New Jersey.

The Institute consists of four schools: Historical Studies, Mathematics, Natural Sciences, and Social Sciences. However, it is not a school in any usual sense. The Institute doesn't offer degrees, and it doesn't offer courses. In fact, the Institute doesn't even have any students. The Institute is a place for deep thought where roughly thirty permanent faculty members in its four schools, together with visiting scholars, pursue knowledge for its own sake.

The Institute opened with five of the world's leading mathematicians and physicists: John Alexander, Albert Einstein, Oswald Veblen, John von Neumann, and Hermann Weyl. Forty of the fifty-six Fields Medalists and thirty-three Nobel laureates have been members of the Institute. The Institute is a place where the world's best theoreticians think deeply about mathematics and physics without worrying about real-world applications.

The Institute is difficult to find; you must seek it out. If you visit the quaint town of Princeton, you will find yourself embraced by the Princeton University community. But you will see no evidence of the Institute; you won't find prominent signs indicating its location. That is intentional. The Institute is a tranquil oasis, set apart from the rest of the world in a nature preserve. It is a haven for brilliant minds. The campus consists of several red brick buildings in the Federal style connected by paved walkways surrounded by expansive fields and forests, and a housing complex for visiting scholars designed by the well-known Bauhaus architect Marcel Breuer.

Fuld Hall at the Institute for Advanced Study (photo courtesy of the Institute for Advanced Study)

My father's professional story has a happy ending, perhaps even a fairytale ending. But his path to professional success is but a part of a larger story, one that raises questions about the price that was paid. My parents never intended to live their lives separated from their families in Japan. Their original plan was a short-term stay in America followed by a return to Japan, where they would raise their family with relatives nearby.

My father made the most of his stay at the Institute. He proved groundbreaking theorems that brought prestigious job offers. The University of Pennsylvania and Johns Hopkins University competed for his services, and Hopkins won out in the end after they sweetened their offer by throwing in a Steinway grand piano. As a result, our family ended up in Lutherville, Maryland, an upper-middle-class suburb of Baltimore, and the prospect of a return to Japan receded into the distance. In view of those lucrative offers, compared with the impoverished postwar conditions in Japan, it is no wonder that they chose to settle in the United States.

For many years, my mother held out hope that the family would one day return to Japan. It is perhaps for this reason that my parents did not become

American citizens until 2007, almost fifty years after they came to America. Obvious challenges accompanied my parents' decision to remain in the United States. They had to learn a new language, and they had to adjust to customs that were at odds with their Japanese upbringing. On top of such formidable challenges, they also had to withstand the weighty emotional and psychological toll of racism.

Life in America in the 1950s and early 1960s for Japanese immigrants was a harrowing gauntlet, in which they suffered abusive insults and stares from some of their former enemies. It was a fresh memory in the American psyche that in 1941, Japan had attacked the United States in a brutal act of war, leading to four years of intense warfare, culminating in the dropping of atomic bombs that destroyed the cities Hiroshima and Nagasaki. At home, in 1942, the American government responded with the forced relocation and internment of over 100,000 Japanese-Americans.

It is then no surprise that there were deep wounds that bred animosity between Japanese immigrants and U.S. citizens after World War II. Such unbridled animosity could be found even among senior government officials. In his testimony before congress in 1943, Lieutenant General John DeWitt proclaimed,

> I don't want any [Japanese] here. They are a dangerous element. There is no way to determine their loyalty ... It makes no difference whether he is an American citizen, he is still a Japanese. American citizenship does not necessarily determine loyalty ... But we must worry about the Japanese all the time until he is wiped off the map.

In a public opinion poll published in 1950 by the Institute of Pacific Relations, thirteen percent of the American public was in favor of the extermination of all Japanese. Perhaps I shouldn't be surprised by this statistic. After all, the U.S. military supported the slaughter of innocent Japanese citizens with their firebombing raids and the nuclear annihilation of two major cities. Those actions saved American lives and brought the war with Japan to an end.

Before 1952, Japanese immigrants were not even allowed to become naturalized U.S. citizens. Japanese-Americans were viewed with suspicion at best, and outright hatred at worst, and that sentiment would take decades to fade away.

My parents arrived in America at a time when Japanese were not welcomed with open arms. On the contrary, they came to an America where their actions and intentions were viewed with suspicion and disapproval. And they were readily identifiable by their Japanese accents and facial characteristics.

My mother told us stories about mundane trips to the neighborhood grocery store in which an attempt to buy bread or eggs was an invitation for verbal abuse. Shoppers would curse them, calling them "Jap" or "Nip," and blame them for war atrocities such as the attack on Pearl Harbor. They were refused service at gas stations by attendants who pretended not to understand their accented English.

Such hostility faded over time, and the public perception of Japanese-Americans improved. Yet they have not vanished completely. In the 1970s and 1980s, anti-Japanese sentiment was displaced onto envy of Japan's industrial success, such as the rise of the Japanese auto industry at the expense of Detroit, as portrayed in the film *Gung Ho*.

In spite of improving attitudes toward Japanese-Americans, there were still instances of outright criminality. I am old enough to remember several hate crimes against our family. While most of our neighbors embraced and welcomed us into their homes, my early childhood in Lutherville, Maryland, in the 1970s was marred by several racially motivated incidents. Once, the teenaged son of neighbors shot out the windows of our kitchen. That same young man left a burning paper bag filled with dog shit on our front porch. During a road trip through Alabama, we were intentionally rear-ended on the highway. The headlights of the attacking driver's car were smashed, and the man forced my parents to pay for the repairs. It was clear that he was enjoying verbally torturing us while we waited helplessly and patiently for the repairs to be completed at a service station.

In the worst incident that I recall, I was the victim. The same wicked neighbor who shot out our windows together with one of his friends committed a hate crime against me, a defenseless second-grader, as I was playing in our backyard. They seized me and dragged me, kicking and screaming, to the forest behind my house. There they tied me to a tree. They pulled the rotting carcass of a dead bird from a paper bag, and they shoved it down my back. I can still see that neighborhood teenager, the one who hated us, clad only in cutoff Levi's and his trademark red baseball cap, as he toyed with his Swiss Army knife while asking me whether I knew what he liked to do to "Nips." Then they left me, tied to the tree, in a pile of dried oak leaves. I was left there in the summer heat for what seemed like hours, sobbing, while flies gathered. My mother finally found me and brought me home.

My parents never called the police, and they didn't confront the culprits or their parents. I thought they were supposed to protect me. I couldn't understand why they preferred to pretend that nothing had happened. Whom were they supposed to protect, me or my wicked neighbors? Why did they choose to do nothing at all?

How often must an impressionable child maliciously be called a "Chink," "Jap," or "Nip" before suffering permanent damage? Although such incidents were infrequent, they are among the most vivid early memories I have from the early 1970s as one of the few Asian kids in my otherwise all-white neighborhood.

I never thought of those extreme incidents as anything more than our poor luck with a few benighted individuals. Now that I reflect back on those events, I wonder what life must have been like for my parents when they first came to America. I have heard of no extreme acts of hatred perpetrated against them, but I am certain that the minor indignities must have been frequent, and the accumulation of such acts was probably more insidious than those few vicious acts that could be chalked up to a small number of fanatical racists. Perhaps that is why my parents' behavior often seemed like paranoid fear.

I now believe that the circumstances I have been describing are part of the reason that I was not taught to speak Japanese at home. I can't even count to fifty in Japanese. My parents always spoke to me in English, and in my presence, they always spoke to each other in English. But in private, they always spoke to each other in Japanese. My parents determined early on that it would be important for my future success to be fluent in English, and they did not press me to learn Japanese. On the contrary, they made no effort whatsoever to teach me to speak Japanese. They wanted me to be an all-American boy for my own good, but they had no idea what that actually meant. They must have been believing and hoping that they could spare me some of the problems they had encountered as Japanese immigrants trying to make a life for themselves in America.

To this day, my parents continue to lock the doors of the house when they are home. They do this out of habit and out of fear. I think it is telling that one of the few Japanese phrases I know is *dorobou hairimasu*, which translates to "hooligans and burglars might enter." I heard this phrase thousands of times as a child, each time my mother discovered an unlocked door.

Kaikin Home

Growing up, I never thought about the challenges my parents faced raising a family under such difficult circumstances. What kid in my position would? World War II was in the distant past, and my parents rarely talked about it.

I didn't know anything about the effects of the war and of that widespread belief in Japan that foreigners, and Americans in particular, were racially inferior. I wonder what my parents thought about the prospect of raising three boys in a country they had been taught to despise.

I knew very little about my family's origins. We had no aunts or uncles or cousins living in America. And we almost never communicated with our relatives in Japan by phone. In fact, I have met my cousins no more than two or three times in all my forty-seven years. My little nuclear family was on its own. We were castaways living thousands of miles away in a foreign country with a foreign culture and language, surrounded by the very people my parents had been brought up to believe were the enemy.

It may surprise you to hear that I first learned of my parents' arranged marriage when I was in middle school. My parents almost never talked about their personal lives, and without ties to extended family, there was no reason for the topic of their marriage to come up, and it never did. Arranged marriages were common in prewar Japan. Bride and groom were selected by friends or family, and, like my parents, they were often strangers when they wed.

As a young *nisei*, I knew next to nothing about being Japanese. For me, being Japanese was all about food, the meals that my mother prepared at home, and my physical appearance, so different from that of my neighbors and schoolfellows—straight black hair, flat face, and slanted eyes. With no Japanese friends, I was constantly aware that I was different, unlike almost everyone I knew.

I was troubled by the fact that we were different from all the other families on our block. I viewed our habit of locking ourselves in as world-class paranoia. Indeed, by the early 1980s, our harassing neighbors had moved, and there was no longer any reason to be afraid.

In recent years, I have changed my opinion. I view the locking of all the doors when we were at home not as paranoia, but as a symbol of my having grown up between two cultures: upper-middle-class America and traditional Japan. I went to school on the outside, in America, while at home, I lived bottled up in isolationist Japan, literally and intentionally locked in to prevent outside influences from entering.

In retrospect, my home life seems a small-scale version of isolationist Japan during the Tokugawa period (1641–1853), when the shoguns, leaders of the military government, enforced a policy called *kaikin*, which largely prohibited contact with foreign countries. The original edict that enforced this policy had seventeen rules, including one that Japanese who secretly attempted to travel abroad were to be executed, and any Japanese residing abroad who returned to Japan were also to be executed.

That isolationism ended in 1853, when Commodore Matthew C. Perry's fleet of ships sailed into Tokyo harbor. The Japanese tried to expel the Americans at first, viewing them, as I mentioned earlier, as "wild beasts," but they eventually acquiesced to the Americans' presence and signed a treaty with the United States in 1854, bringing an end to the shoguns' *kaikin*. Although isolationism vanished as an official policy, its underlying principles lived on in the ideologies of subsequent Japanese leadership, and it was deeply ingrained in Japanese culture when my parents were growing up. Some of its effects still reside in them.

Given this culturally pervasive idea of Japanese superiority and the anti-Japanese racism that my parents experienced in America, it makes sense to me that my parents would, perhaps unconsciously, enforce a kind of *kaikin* at home in quest of self-preservation and protection for themselves and their children.

They had no idea that their actions presented us children with challenges in our lives on the outside, our lives in America. Why would they? My mother didn't interact with many Americans. Apart from her housewifely errands, she spent all of her time occupied with housework and the raising of three boys. My father was completely engrossed by his mathematics. They knew next to nothing about the country they lived in, as if they were tourists in a foreign country who never ventured beyond the gates of their all-inclusive resort.

As a kid with no knowledge of my family's history, I had no way of knowing all of the forces at play in their parenting. I could only compare our family with the others in the neighborhood, and it seemed to me and my brothers that we had gotten a raw deal.

Chapter 3

$\sim$

MY CHILDHOOD (1970–1984)

*L*utherville, Maryland, is a hilly network of twisted streets dotted with nearly identical 1960s-era single-family split-level homes. Our neighborhood was home to baseball legends Mark Belanger, Jim Palmer, and Brooks Robinson of the Baltimore Orioles. I loved the Orioles. Their victory over the Philadelphia Phillies in the 1983 World Series is one of the most cherished memories of my life in Lutherville. Little else from my childhood resembles anything that would have been considered normal in my neighborhood.

Ken-chan, Lutherville, Maryland (1970s)

When no one was practicing music, our house was silent. You could hear the compressor of the fridge unless the dishwasher was running, and once in a while, you would hear the scraping of Igor the dachshund's paws as he wandered about, or the quiet thud of my mother's slippers as she did chores. All five of us were probably at home—my mother at her chores, we boys in our bedrooms, and my father in his study doing math all day. We always had to be quiet so as not to disturb my father's concentration.

My mother, Sachiko, was raised to be a model Japanese housewife, invisibly running the household in the background. Her education included such refinements as flower arrangement and meal presentation, and for a brief period in America, those skills were channeled into portrait painting. She has devoted her entire adult life to Takasan and her three sons.

I know very little about my mother's childhood; she never talked about it. What I do know is that she ran our home like clockwork. She is the most dependable person I have ever known, and she never let time get the better of

© Springer International Publishing Switzerland 2016
K. Ono, A.D. Aczel, *My Search for Ramanujan*, DOI 10.1007/978-3-319-25568-2_3

her. For example, even though a permission slip for a school field trip might be due in two weeks, she would nevertheless sign it the day she received it. She would pay bills on the day she received them out of fear that something might go wrong if she waited a few days.

She took her role as a traditional Japanese mother very seriously. She cooked our meals, did our laundry, and cleaned the house. She performed those chores almost single-handedly. None of us boys ever had any chores. Every morning, without fail, I awoke to find a delicious breakfast waiting for me on the kitchen table. Although my father worked late and slept in, his breakfast was also already nicely set out for him bright and early, a meal that he would enjoy two or three hours later. My mother ate her breakfast alone after we boys left for school. After school, I came home to find that my bed had been neatly made, as were all the beds. My mother's habits of waking up hours before everyone else and preparing breakfast and of making all of our beds is indicative of her commitment to the family's affairs.

The dinners she cooked consisted of delicious Japanese fare, such as chicken tonkatsu, shrimp tempura, or beef curry, which appeared like clockwork at 6:00 p.m. every day. We would all gather as a family for dinner, often the first and only time any of us would have seen my father all day. After dinner, he would retreat to his office and return to whatever mathematical problem was on his mind.

Taking care of the home and the family's affairs was my mother's full-time job. She took her role so seriously that she never had time for much of anything else. In a good year, she might go out to the movies once or twice. She almost never watched TV. We almost never went out to eat, because it was her job to feed the family. I have just a couple of memories of her going on an outing with friends. Apart from her hobby of copying famous paintings, which she pursued only when I was young, she didn't seem to participate in any activities other than housework and parenting.

She rarely bought new clothes, and she never went to the beauty salon. Despite the fact that she had been close to her sisters, she rarely phoned them in Japan. The long-distance charges were too expensive. She spoke with them at most once or twice a year.

I can't imagine how she managed to live such a life. It would have driven me mad. It seemed to me that my mother had no identity and independence apart from the family. For her, serving the family was her job, her duty.

But it annoyed me terribly that she constantly reminded us how much she was giving up. She presented herself as a martyr who had sacrificed all self-interest for the family. I think that this was perhaps her way of instilling in us a sense of duty to succeed in the lives that they had planned for us. I certainly didn't want to feel guilt for not doing my part. But I thought that she should enjoy a better life and do something for herself once in a while, like all the other mothers in the neighborhood.

For twenty years, I also misunderstood my father's role. I thought that his disinterest in me was somehow a reflection of his low opinion of me. By comparison, my friends had very different relationships with their fathers, loving relationships. They would play catch in the yard, go for bike rides, attend baseball games, and do other normal things. I could see it all from my house, from behind the windows of my bedroom.

As the only Japanese-American family in our "All-American" neighborhood, I had no way of knowing that I was trapped at the confusing and frustrating intersection of incompatible cultures. I understand my family dynamics much better now. In traditional Japanese families, it is quite common for fathers to be almost entirely absent and removed from their children's lives. The fact is that Japanese fathers are expected to spend so much time at work that they often have little time and energy to spend with their children. Japanese companies place heavy demands on their employees, and as a result, the culture has adapted by placing the responsibility for raising children and overseeing education almost entirely on the mothers. The Japanese father is the breadwinner; it is his duty to provide for the family.

Despite the fact that the father spends little time at home, he is highly regarded and respected by all members of the family. Children's positive views of their fathers are often the result of their mother's efforts to portray their husbands as someone to be respected and revered. Japanese fathers are often held up as role models for their sons.

This makes sense to me now. I was indeed brought up to revere my father as the world-class mathematical genius that he is. Everything we had, we credited to the importance of his work. And as a theoretical mathematician, most of his work was done at home, closeted away in his study on the first floor of our house. He would spend hours on end in his office, behind a closed door, scribbling crazy-looking formulas on yellow pads of paper. That was his duty. When I was discussing this recently with my mother, she said, "Ninety-nine percent of his life has been devoted to mathematics, with one percent spared for his family and hobbies. Make sure to tell people."

This may seem shocking to Westerners, but from the viewpoint of traditional Japan, it makes sense. As a kid living in America, it made no sense to me. Over time, I would conclude that the whole unsatisfactory situation was my fault, that I had done something wrong, and my father had rejected me.

While my brothers had other gifts, it was clear that my gift was mathematics. And as tradition would have it, this meant that my father was to be the role model for his budding mathematician son. Although my relationship with my father was emotionally detached by Western standards, he was actively involved in my mathematical education when I was a young boy. All of my cherished memories from early childhood are related to the bond that mathematics forged between us.

My parents like to tell the story of how I "discovered," at the age of three, that there are infinitely many numbers. I argued that if it were not so, then there would have to be a largest number. But that makes no sense, because one could always add 1 to that largest number to obtain an even larger number. Therefore, there must be infinitely many numbers.

I have fond memories of sitting at a little kiddie desk in my father's office working on geometry problems while he, scribbling at his large steel desk, worked on his grownup math problems. My father taught me beautiful theorems by making use of the fun I had in calculating large numbers. When I was still in elementary school, he taught me Fermat's little theorem in this way, and he presented it to me like a magic trick. It went something like this.

"Ken-chan, pick a number that isn't a multiple of 7." I'd pick a number, such as 13. "Now raise it to the sixth power and subtract 1." I'd run around the corner of his giant steel desk, to my little kiddie desk, and I'd work it out by hand. After a minute or so, I'd arrive at the answer: 4,826,808. "I bet that it is a multiple of 7." I'll take that bet, I thought. He has only a 1 in 7 chance of being right, and I'll take those odds any day. And I'd run back to my kiddie desk, and I would divide 7 into this large number, and I found that he was right: $4,826,808 = 7 \times 689,544$. I'd say something like, "You got lucky. Let's do it again."

"Ken-chan, pick another number that isn't a multiple of 7. Raise it to the sixth power and subtract 1. I bet the number you get is a multiple of 7." Now I knew there had to be a catch, because this time he knew the answer even before I picked my number. I'd then pick a number anyway to humor him, like 29, and I'd run around the big steel desk to my little kiddie desk, and after a few minutes, I'd figure out that $29^6 - 1 = 594,823,320 = 7 \times 84,974,760$, a multiple of 7, as expected.

I then asked him why the number 7 is so special. And he'd respond by picking another magic number. "Ken-chan, pick any number that isn't a multiple of 13. Raise it to the twelfth power and subtract 1." I'd immediately be able to guess what came next. But I'd let him say it anyway. "I bet the number you get is a multiple of 13." I liked manipulating large numbers, but not the supersized ones that you get by raising to the twelfth power. Sensing that I'd figured out the rule, he then told me Fermat's little theorem: If p is a prime number and a is not a multiple of p, then $a^{p-1} - 1$ is always a multiple of p. Then he explained a proof. After all, what we had done only offered evidence of the assertion. For it to be categorized as a theorem, an undeniable truth, it needed a watertight explanation, a proof, that showed that the statement holds for every prime number p, not just 7 and 13, and for every number a that is not a multiple of p.

My father enjoyed telling me stories about great mathematicians. I learned about Euclid, Euler, Descartes, Fermat, Galois, Gauss, Weil, and others. These giants had written great books, and it seemed that we had all of them. I loved holding those old books, flipping through the pages, and enjoying their sweet vanilla scent. Those books were filled with strange symbols and Greek letters like τ (tau) and ϕ (phi) that seemed freighted with a significance that I could only guess at. Although I couldn't understand a word of those books, I dreamt that one day I would be able understand them, and perhaps even write one myself.

My most vivid memory from that time is of my father playing a special cassette tape on his boombox, a static-filled recording of David Hilbert giving a talk in the 1930s. Hilbert is generally regarded as one of the most influential mathematicians of the nineteenth and twentieth centuries.

He was the leader of the Göttingen collective of mathematicians. His students and collaborators included many famous mathematicians, including Alonzo Church, Erich Hecke, Emmy Noether, Hermann Weyl, Ernst Zermelo, and John von Neumann. At the 1900 International Congress of Mathematicians, Hilbert put forth a list of twenty-three of the most important unsolved problems in mathematics. That list served as a guide for much of twentieth-century mathematical research. It is considered the most successful and influential list of problems ever assembled. Mathematicians spent the last century working on those problems, and along the way, they developed an extraordinarily rich trove of important mathematical ideas and techniques.

My father treasured the Hilbert audiotape; it remains one of his prized possessions. It was exciting to watch my father as he listened to Hilbert's lecture, even though I didn't understand a single word. He'd close his eyes as he listened, doing his best, I supposed, to imagine himself in the lecture hall. For my father, the tape represented an important historical event, one that he felt was as momentous as the famous 1877 phonograph recording of Thomas Edison reciting the nursery rhyme *Mary Had a Little Lamb*.

The role of mathematics in my early childhood was not confined to our home. I was identified by Julian Stanley, a famous psychologist at Johns Hopkins, as an uncommonly talented kid. Stanley was an advocate for the accelerated education of academically gifted children. I was one of the children he studied in his well-known "Study of Mathematically Precocious Youth," a longitudinal survey of children who scored 700 or higher on the SAT math test before the age of thirteen—thereby putting them among the top one percent of those taking the test. Enrolled in his study, I took the SAT several times before the end of middle school. I was in sixth grade when I scored over 700 on the SAT math section on my first try. Thanks to that study, I was offered a scholarship, which I declined, to attend Towson State University the following year.

That is how my life went in elementary school. My mother handled the day-to-day parenting, and my father, while emotionally detached and consumed with his mathematics, found a way to offer me love and affection through our common bond of mathematics. I excelled in elementary school, and I enjoyed my notoriety as the smart math kid.

Books or Bikes, Lutherville (1976–1984)

Despite my fond memories of working with my father on mathematics at my little kiddie desk, I actually never wanted to be a mathematician. That wasn't the life I wanted; it was the life that others had worked out for me. I had enjoyed the label as a gifted math student in elementary school, but as I got older, I wanted to be more like all the other kids in the neighborhood, and they weren't doing math for fun with their fathers. I just wanted to hang out, play baseball, and go to the movies.

As a middle-school pupil, the numbers I appreciated were to be found not in mathematics books, but on baseball cards. I was a dedicated collector of Topps cards, and I put my math skills to use in studying the statistics about the players depicted on those cards. To this day, I remember a bunch of odd baseball statistics. For example, the number 388 reminds me of the cherubic face of Rod Carew, the Minnesota Twin who compiled a league-leading batting average of .388 in 1977, the year he magically approached the elusive .400 barrier, which has not been breached since Hall of Famer Ted Williams hit .406 in 1941.

I dreamt of a life as a NASA astronaut. In middle school, I collected glossy pictures of Apollo moon missions, writing dozens of letters to NASA requesting the photos. Carl Sagan, the creator and star of the TV show *Cosmos*, was my hero. I read his books, even the ones I had no chance of understanding. I was very good at math, and I was confident that this meant that I could become a scientist like Carl Sagan.

Around seventh grade, an awkward age for most boys, I began, like other boys, to struggle with my identity. Everyone's struggle is unique; this was mine:

A late bloomer, I was the shortest kid in class. I wore glasses, which was not fashionable in 1980. I was the only Asian kid in my class. That was already three strikes against me. But above all, without the maturity to understand that I was trapped between two cultures, the negative parental voices in my head started becoming more and more insistent:

Ken-chan, you small and weak… You no fit in… You must be best math student… You cannot afford people think you stupid… There is no other way.

Although my father was largely absent as a parent, I was aware of his opinions and concerns regarding me. I learned about them from the conversations I overheard between my parents. My parents frequently talked about us in private, and in our house, which had very thin walls and doors, very few conversations were actually private. Of course, I didn't understand what they were saying, since they were speaking Japanese, but I heard my name, so I understood that they were talking about me, and from their intonation, I could infer that they were worried.

When my mother would speak with me following such a conversation, I was annoyed and confused, for it seemed as though my mother was merely the messenger for a higher power. When she criticized me, I always wondered whether she was expressing her own opinion, or whether perhaps she did not really think so disparagingly of me and was simply repeating the edicts handed down by my father. I felt his power over me as an invisible and omnipotent judge, and I began to crave his praise even more for my accomplishments, which however good they might be, were never good enough.

When my father occasionally descended from the clouds for some hands-on parenting, it usually meant that I had done something that merited punishment. If he had to take time from his important mathematical research, then the stakes had to be quite high, and I had to be ready for substantial consequences. Such incidents were rare, but I recall one time when he beat me severely.

When I was twelve years old and in sixth grade at the Hampton Elementary School, I had the privilege of helping out as an office assistant. I was allowed to lend a hand in preparing announcements and to help out in other ways in the school office. The office was a suite of three rooms whose entryway opened into a waiting room where the receptionist had her desk. Connected to that room was the principal's office, which I thought of as a torture chamber: kids went there only when they were in trouble. Then there was the copy room, a debris field of slotted mailboxes, stacks of reams of paper, large rolls of construction paper, and a spirit duplicator, generally known as a Ditto machine.

In the late 1970s, Ditto machines were enjoying their last hurrah before the xerographic copiers that we know today took over. The Ditto machine was a low-volume duplicator that used alcohol-based inks and a "master original" that could be either handwritten or typewritten. The master was a two-ply affair whose top sheet was the one that was written or typed on, and whose second sheet was covered in a layer of wax impregnated with ink, usually purple. After the top sheet had been written or typed on, it was removed, and the waxy sheet, which now contained an imprint of what had been written on the top sheet, would be placed on the drum in the printer. Each turn of a crank produced a copy. Compared to the photocopiers of today, which spew out limitless copies at the push of a button, those Ditto machines were a pain, and they were messy if you didn't know what you were doing.

The principal of Hampton, Mrs. May C. Robinson, was a strict and awe-inspiring executive. That spring, Hampton came under threat of closure. School budgets were tight, and closing Hampton was proposed as a way to save money. Mrs. Robinson leaped to the defense of her school. She prepared a carefully handwritten letter on a master ditto that asked parents to support the school by protesting the proposed closure. She requested that they sign the letter and return it to the school to show their support for her petition to keep Hampton open. Alone in the copy room, I discovered the spirit master, and in an act of stupid mischief whose obvious consequences I somehow failed to foresee, I vandalized the letter. On the line intended for the parents' signatures, I wrote "GLORY! Close the fucking school."

Actually, I loved Hampton. I was simply playing an incredibly stupid preadolescent prank. The secretary didn't notice what I had done. She innocently duplicated the letter and then distributed the flyers to the teachers. Some of the teachers sent the flyers home with their pupils before someone noticed what had been perpetrated.

The next day, I was busted. I readily admitted to my stupid stunt, and I was prepared to accept my punishment. I was sentenced to a loss of office privileges and the stool in the corner of Mrs. Robinson's office, one hour each Friday for the rest of the school year. Her office had glass walls, and so I prepared for the abject humiliation of being stared at and laughed at by all the kids in school as they made their way to the lunchroom.

Despite my admission, my parents refused to believe that I had pulled the stunt, and they argued with Mrs. Robinson that I was incapable of such an act, going so far as trying to prove that I didn't know any curse words. They suggested that another student must somehow have coerced me into taking the blame, despite the fact that the handwriting was clearly mine. They finally overcame their cognitive dissonance and came to understand that I was the guilty party. Embarrassed that they had come to my defense, my father scolded me, and my reward was the severe spanking that I mentioned above.

My parents had come to my defense because they thought that they knew exactly who I was and what I had in me. They "knew" that I was incapable of such misbehavior, which was so remote from the formula that they had established that dictated my life. Perhaps my prank was a cry for help, an attempt to show my parents that I was something more than the x that had to satisfy their equation. If it was such a cry, it certainly failed in its purpose. I received my spanking, and life went on as before.

But those formulas required success, and we boys had to do our part to meet those requirements. My parents believed that Momoro and I had the talent to be "the best," perhaps as an artifact of their upbringing, which promoted the idea that the Japanese are a superior race. Perhaps they simply couldn't accept the idea of their children being anything less than the top. They believed that Momoro could become a world-class pianist, another Vladimir Horowitz, and that I could become an influential mathematician.

Like any child, I wanted my parents to be proud of me, but if they were, I never knew it, for their philosophy of parenting left no room for praise. I was especially sure my father, who played such a small role in day-to-day parenting, considered me a failure. I had no idea that traditional Japanese families were different from the American families that were all I had for comparison. How was I to know? And if I had known, would it have helped? I craved my father's approval, a need that would haunt me for decades.

My parents deprived me of almost everything that didn't have an academic purpose. As I have mentioned, we were the only Asian-American family in our neighborhood, and so I expected that there would perhaps be some differences between my family and those of my schoolmates. But I couldn't come to terms with this degree of isolation and restriction. It seemed a cruel fate. As I got older, I realized that I had never played catch with my father in the yard. I had never been allowed to go to a sleepover at a friend's house. I wasn't allowed to hang out with friends on weekend nights. I wasn't allowed to have anything that was part of a normal American childhood.

This isolationism also applied to our nonexistent spiritual lives. Although my mother had shown some interest in joining a church in search of some sort of community, we had no religious affiliation. My father saw no point to it, and so my mother was easily outvoted. He didn't believe in God. He didn't believe in anything he couldn't see or prove. Everything about his worldview depended on his strong belief that there are no mysteries or sources of wonder in the world that couldn't be figured out with the help of the scientific method and mathematical logic.

My father joked that the only reason to go to church was as an insurance policy in the unlikely event that he happened to be wrong about the nonexistence of God and an afterlife that would be denied to heretics like him. He could understand why people wouldn't want to miss out if the afterlife were pleasant. But that was not reason enough for him to waste time in church, and as a result, I was raised an agnostic.

I had no reason to question my father. After all, he was the man who had conjured a deep and important theory about objects called "algebraic tori" and "Tamagawa numbers." It seems that he had solved a problem that world experts had considered unsolvable. If there was a divine figure for the Ono boys, it was Takasan, our math genius father.

As I reflect on my childhood from the vantage point of middle age, I have come to believe that my parents became cynical about religion as a result of the defeat of their spiritual government in World War II, when Emperor Hirohito was forced to surrender and give up his divine status. What would it take after that for my parents to open themselves to the possibility of a life of the spirit?

I was thus never exposed to the idea that we are all are part of a larger community, each with the responsibility of trying to make the world a better place. Instead, I grew up to be a cynical adolescent who viewed humanity as a group of people mindlessly racing and competing with each other for their rightful place. I saw little beauty in the world, and I certainly didn't recognize anything that inspired awe and wonder. I almost had no soul. I was body and brain and no heart. It would be many years before I would recognize how sad such an existence was.

As a freshman at Towson High School, a school that would later claim Olympic swimming champion Michael Phelps among its alumni, I lost my way and my identity. My transition from middle school to high school destroyed the little connection I had with life as a teenager living in America. Compared to my middle school, Ridgely Junior High, Towson High was huge, and to me, its students were huge. It was a soulless place that just swallowed you up and spit you out four years later. I had walked to Ridgely, a freedom that offered me the opportunity to hang out with friends both before and after school. But I rode the

ANNALS OF MATHEMATICS
Vol. 78, No. 1, July, 1963
Printed in Japan

ON THE TAMAGAWA NUMBER OF ALGEBRAIC TORI

BY TAKASHI ONO

(Received April 30, 1962)

TABLE OF CONTENTS

My father's famous paper in the *Annals of Mathematics*

bus to Towson, and due to the way the school boundaries were set, almost all of my middle-school friends ended up going to other high schools. The little freedom I had before was now gone.

With college around the corner, my parents stepped up their tiger parenting. My life was reduced to a mindless pursuit of first-rate academic credentials. Any score other than a perfect 800 on that SAT math test would have been a terrible embarrassment to the family. Any grade below an A would have been a personal disaster. A class ranking outside the top five would have meant the end of the world.

As a sophomore, I also struggled with my identity as one of the few Asians in school. I understood that I had a physical appearance that was different from that of most of my classmates and that as the child of immigrants, there were some cultural differences to be expected between my family and theirs. But it was too much for me on top of the frustrating and confusing life as the child of tiger parents to have to deal with hurtful racially motivated teasing. Even though I understood that none of it was malicious, I still couldn't stand it. I hated being called "Jap" or "Nip." I hated hearing "Oh no, there's Ono." I simply wasn't strong enough to embrace my heritage. I wanted to be accepted, and the teasing made me feel isolated, different. It would be many years before I could live comfortably with my Japanese heritage. At the time, I was simply embarrassed. Why did I have to have parents with strong foreign accents? Why did I have to be the kid with the strange last name? Why did I have to be the straight-A student whose parents expected and demanded more? What did I have to do to finally earn love and respect from my father? I desperately wished to be someone else, almost anyone else. I had become a pitiful loner who walked the halls of school head down, wishing I was invisible.

Those feelings were compounded by my having been labeled a "gifted and talented" student, one of a tiny cohort of kids identified by the Baltimore County Public Schools for our ability to ace standardized tests. We took all of our classes together, and we marched from class to class as a group as if under quarantine. My social role in school was clear. I was the archetypal Asian-American nerd. Being marked out as gifted was no comfort to me, nor was I comforted by the fact that my classes were easy for me. I hated almost everything about high school.

I was being crushed under the high expectations set for me at home, which had always been more crucible than nest but now was unbearably so. How could it have been otherwise? My parents owed all of their success to my father's prowess as a mathematician. Their few friends were other professors with overachieving children who were being accepted by top private colleges and winning elite music competitions. The way my parents raved about them, those professors' kids were models of perfection. How could I even think about measuring up? I had to get into a college like Princeton or MIT; the local state university would have been a humiliation. It was all or nothing. The pressure of trying to live up to those paragons was overwhelming.

Under those suffocating circumstances, I lost interest in school. I lost hope that I could amount to much of anything. I had been taught that every worthy scientist solved problems that nobody before them could crack. I had learned that accomplished young scientists must earn the endorsement of a master scientist, someone to do for them what André Weil had done for my father and other young Japanese mathematicians in the 1950s. I had learned that one could reach the highest level of honor and respect by overcoming formidable obstacles, such as isolation and poverty.

And here I was with no obstacles, no excuses, yet no chance of reaching any level of honor or respect. I wasn't starving. I wasn't subsisting on meager bowls of rice while trying to stay alive in a world of smoke and fire. And yet I didn't have a prayer of replicating my father's formula for success through mathematics. But I had no choice. It was the path that had been laid down for me. I had been brainwashed into believing that there was no other way for me to have a happy life. It was actually worse than that: I had been brainwashed into not even thinking about what might constitute a happy life. It never occurred to me that I might step outside my preordained formula into something of my own choosing.

Yet I somehow realized that I was going to go mad if I didn't find something positive in my life. I needed an escape, something outside of grades and exams to live for. In seventh grade, I discovered cycling. When I was on my bike, life became simpler, with all its complications reduced to pedaling, watching where I was going, and feeling the liberating wind in my face. Two years later, halfway through ninth grade, cycling became an addiction. I took up bike racing, and soon nothing else in my life seemed to matter. At first, I belonged to a small recreational racing club. But the following year, emboldened by my success in defying my parents by quitting the violin, I audaciously stepped up my training and joined the best team in the area. My will was so strong that there was nothing my parents could have done to dissuade me from pursuing my new passion. Of course, they tried to make me give it up:

> *Ken-chan, Good students no have time for bicycle riding… It dangerous. You could get killed… It take too much time… So much exercise kill you… You will get so tan that you will look like* obake, *a zombie.*

I listened, but I refused to yield. I lived for the thrill of hurtling down steep hills in a tight aerodynamic tuck. I loved the harmony that I experienced with my machine when I climbed long steep hills out of the saddle, as if we were in a spirited dance. I cherished the mental freedom that I felt after pedaling miles and miles over the rolling hills that populate the landscape north of Baltimore. In my mind, I raced famous cyclists, such as Eddy Merckx and the budding American star Greg Lemond. I did all this to drown out the painful voices that haunted me when I wasn't riding.

To pay for this expensive sport, I worked part-time at Valley Pharmacy as a stock boy earning minimum wage. I swept the floors, stocked the soda machine, and delivered prescriptions to customers, one of whom, to my surprise and delight, was Johnny Unitas, the retired Hall of Fame quarterback who had played for the Baltimore Colts from 1956 to 1972.

I lived for cycling, and much like David Stohler, the teenage protagonist in *Breaking Away,* the 1979 film about a cycling-obsessed boy coming of age, I was at odds with a father who didn't understand me.

I raced for the Charm City Velo Team, an elite group of local bike racers, alongside my high-school friends David Lanham and Greg Asner. To my parents, such friends were anathema, kids who would never amount to much, and they were never allowed to enter our house. My parents' predictions notwithstanding, both of these cycling buddies would achieve success. David has enjoyed a distinguished career in the military, and Greg is a well-known environmental scientist at Stanford.

We had sharp-looking team kits that included red-and-green skintight Lycra jerseys. We wore sleek sunglasses, and I was the first to get a pair of supercool Oakleys. The first models were more like face shields, and they made me look like Darth Vader, which was fine with me, since I hoped to inspire fear in my racing opponents.

Goofing off before a race

Several of our older teammates were college students at Johns Hopkins. When I first joined the team, I had the naive hope that my parents would somehow approve of cycling because I was in the company of Hopkins students. I hoped that they would approve of these friends and somehow support me in my chosen sport. Those hopes vanished quickly. Neither of my parents ever uttered a positive word about cycling. They never attended any of my races. They never even asked about them. They probably had no idea, if they even thought about it, where I had been, for we raced as far away as Pennsylvania and Virginia.

My friend Peter Verheyen, who would attend my wedding in 1990, was one of my teammates who understood me and sympathized with my situation. He was a strong student at Hopkins, a top regional bike racer, and he was also the son of a Hopkins professor. He was more than a teammate; he became a confidant and older brother. Like my seventh-grade teacher Mrs. Sprankle, Peter gave me much-needed nurturing.

Pat Liu, a biology major at Hopkins who earned a medical degree from the University of Pennsylvania, was our team's star sprinter. In 1983, Pat won the National Capital Open, an important bicycle race held each year in Washington, D.C., on the Ellipse, the one-kilometer-long street in President's Park between the Washington Monument and the White House. I enjoyed drafting behind Pat on our training rides, because his large frame created a nice slipstream that reduced my workload, and I was mesmerized by the sight of the veins in his calves that popped out like pencils when he pushed hard against the pedals.

I lived for our weekend group training rides and road trips, and I enjoyed the fact that our team was really good. My cycling friends became my source of strength and self-esteem. Bike racing kept me sane by providing an escape from my isolationist home and the burden of impossible parental expectations.

On my long solo training rides in the spring of 1984, I plotted my escape, my metaphorical *seppuku*, the ancient Japanese ritual suicide practiced by samurai who had brought shame upon themselves and their families. I had decided to drop out of high school and leave Lutherville in order to escape my hopeless circumstances. I would start life anew, somewhere, somehow, consigning my former life to a black hole of forgetfulness. Although I didn't have a concrete plan, there was no way that I was going to graduate from Towson High School. Nothing would change my mind about that, even if it meant having to live on the streets.

Ono Family mid 1980s (*left to right*: Takasan, Ken, Santa, Momoro, my mother with Igor)

I tried to convince my mother that dropping out was a good decision. I argued that staying in high school was a waste of time. The classes were too easy, I said, and I hated them. I also argued that I could get into a top college without a high-school diploma. My mother, however, maintained that it was an absurd idea, and she fought with me for months:

> *Ken-chan, if you drop out then never come back to this house… You will disgrace family, even worse than now with your tan face and long hair… Why you want punish us after all we done?*

I never explained the real reason for wanting to drop out: I was suffocating. I was desperately seeking freedom and independence. I couldn't take the daily barrages of criticism without any hope of meeting my parents' expectations. I had reached a breaking point.

Chapter 4

~

AN UNEXPECTED LETTER

Lutherville, Maryland (April 7, 1984)

*M*y legs are on fire. I'm fighting to spin the pedals of my svelte French Peugeot racing bicycle in a desperate effort to keep pace with Belgian cycling champion Eddy Merckx. We're battling mano a mano, racing up the switchbacks of Mont Ventoux, the "Giant of Provence," a cruel peak that had claimed the life of British champion Tom Simpson during the 1967 Tour de France. We're racing for the finish line, a strip of white paint in the steep road at the summit, a desolate place marked by a decaying weather station. Merckx, known as the "Cannibal" because of his insatiable appetite for victories, sets an infernal pace. Somehow, I'm able to keep up, while one by one, all the others have fallen behind. The finish is finally in sight, and the fans are in a frenzy. Despite the overwhelming pain and self-doubt building inside me, I summon all my remaining strength. I rise out of the saddle and swing my Peugeot side to side in a furious sprint. And wondrous to relate, I leave the Cannibal in my wake.

I am in a trance. I am not actually in Provence. In reality, I am training for next weekend's National Capital Open, a race that Pat won last year with a ferocious sprint. My solo training rides are mental and physical adventures. They are my first experiments in deep meditation and visualization. I will eventually learn that I am at my best in a trancelike state, where I enjoy a union of body, mind, and soul.

It's a gorgeous brisk Saturday morning, April 7, 1984, and I have been riding my fancy French racing bicycle in the picturesque rolling countryside north of Baltimore among attractive manicured horse farms. Clad in wool cycling shorts, a yellow jersey, cleated Italian cycling shoes, and a flimsy hairnet helmet, I am a sixteen-year-old Japanese-American version of the kid David Stohler from the movie *Breaking Away*.

© Springer International Publishing Switzerland 2016
K. Ono, A.D. Aczel, *My Search for Ramanujan*, DOI 10.1007/978-3-319-25568-2_4

I eventually pull into our driveway on Welford Road and dismount. I walk to the mailbox; I hobble, actually, on account of my cleated cycling shoes, which emit a jaunty clip-clop on the sidewalk. I open the mailbox, and inside, I see a delicate yellowed rice-paper envelope covered with exotic stamps. It is addressed to my father, Professor Takashi Ono. That letter is going to change my life, though it will take me ten years to understand its message.

Exhausted from my ride, I blink to clear away the beads of sweat that are dripping from my eyebrows. This allows me to make out the strange letter. The stamps tell me that the letter is from India, and the postmark makes it precise: it is from Madras. With the curious letter in hand, I ring the front doorbell so that my mother will know to unlock the steel front door and the outside screen door to let me into our crucible of a home.

My two older brothers no longer live at home. Momoro is now attending the Juilliard School of Music on scholarship. Santa is also away; he is a senior at the University of Chicago, where he is just about to graduate with a degree in biology. He will repudiate his predetermined path and outgrow his role as the black sheep of the family by earning faculty positions at the Harvard and Johns Hopkins Medical Schools. He will later be named the twenty-eighth president of the University of Cincinnati.

I am much younger than both of my brothers, and so I have essentially been brought up in their wake as an only child, single-handedly by my mother. Momoro left home at an early age to attend the Curtis Institute of Music in Philadelphia, a boarding school for gifted musicians, and I have few memories of life with him at home.

Still in my cycling shoes, I hobble down the flight of stairs to my father's office, where he is busy, as always, scribbling mathematical formulas on yellow pads of paper while listening to a transistor radio softly playing classical music. I hand him the curious envelope without a second thought.

After lunch, my father calls me into his library, my former bedroom, which is now covered wall to wall with math books and is dominated by a large refectory table stacked high with papers and books. He is holding the letter from India in his hand, and although he is ordinarily a stoic and almost emotionless man, I can tell by the look on his face that he has been deeply moved by the letter. I'm not sure, but I get the sense that there are tears in his eyes.

The letter is typewritten on delicate rice paper, and the letterhead features a rust-colored sketch of a serious-looking Indian man whose thick hair is parted, Western style, on the left side.

For my father to take even a few minutes away from his mathematics, this letter must be important. "Ken-chan, I have to tell you an amazing story about this letter."

S. Janaki Ammal
W/o (Late) Srinivasa Ramanujam
(Mathematical Genius)

S. Ramanujan
1887-1920

45, Muthiah Mudali IInd Street,
Krishnampet, Madras-600 014.

INDIA

Date 17·3·1984

Dear Sir,

I understand from Mr. Richard Askey, Wisconsin, U.S.A.,
that you have contributed for the sculpture in memory
of my late husband Mr. Srinivasa Ramanujan. I am happy
over this event.

I thank you very much for your good gesture and wish
you success in all your endeavours.

Yours faithfully,

S. Janaki Ammal

Janaki Ammal's letter

Dear Sir,

I understand from Mr. Richard Askey, Wisconsin, U.S.A., that you have contributed for the sculpture in memory of my late husband Mr. Srinivasa Ramanujan. I am happy over this event.

I thank you very much for your good gesture and wish you success in all your endeavours.

Yours faithfully,

S. Janaki Ammal

My father explains to me what the letter is about. The letter is from Janaki Ammal, a destitute Indian woman in her eighties who lives in Madras (now known as Chennai). She thanks my father for his gift, a donation that helped fund the commissioning of a bust of her late husband, Srinivasa Ramanujan, a man who had died in 1920.

Despite living in near poverty for over sixty years, rejected by her husband's family, the forgotten brokenhearted widow had only one request when a reporter found her living in a Madras slum in the early 1980s. She had been promised a statue to honor her husband at the time of his death. The promise had not been kept. She desperately wanted a statue erected to honor her husband's memory.

Janaki Ammal in 1987

After reading about this in an Indian newspaper, Dick Askey, who would later become one of my colleagues at the University of Wisconsin, solicited donations from the mathematicians of the world in a massive fundraising campaign. He used the proceeds to commission sculptor Paul Granlund to fashion a bust of Ramanujan. Askey came through and fulfilled the broken promise, and my father was one of the many mathematicians who made a contribution.

Granlund's bust
of Ramanujan

I ask, "Who's this guy Ramanujan? What did he do?"

My father tells me the most incredible story, about the life of Srinivasa Ramanujan. It is the story of an Indian man who overcame incredible odds to become one of the most romantic and influential figures in the history of mathematics. It is the story of a self-taught dropout whose ideas came to him as visions from a goddess. It is the story of a man who had the courage to send his ideas to random mathematicians at the University of Cambridge, and then to accept the invitation of a world-class mathematician who recognized his genius and travel halfway around the world to work with him in England. It is the story of a man who suffered racial prejudice as he strove for accomplishment and recognition. It is a story of a man who would then die tragically at the young age of thirty-two.

I am stunned by this tender story. I am also surprised that this short letter, which is really no more than a form letter, has stirred up such deep emotions in my father, a man I thought had feelings only for numbers and formulas.

However, I recognize at once that Ramanujan's biography mirrors my father's life in many ways. Both men are self-taught creative geniuses. Both men escape poverty thanks to the generosity of a world-class mathematician who offers the opportunity to work with the world's best in a foreign land. And both men are rewarded for their achievements despite the indignities and hindrances to success that they suffer due to racial prejudice.

Chapter 5

~

MY ESCAPE

Ramanujan's story, as told to me by my father, offered me, a teenager about to drop out of high school to escape a frustrating and confusing life, hope that like him, I could perhaps accomplish something in my life. I was encouraged by the fact that the dropout had achieved not only success but greatness, and I was made hopeful by the fact that the dropout's parents continued to support him despite his troubles. Most of all, I was stunned that my father held this college dropout in such high regard; he revered this man as some kind of demigod. I found that odd, because my father didn't believe in anything that he couldn't see or prove, and Ramanujan seems to have had a significant mystical side to his personality. I wondered how he reconciled Ramanujan's claim of inspiration through visions of the goddess Namagiri with his personal beliefs.

Although I didn't have much confidence that I would ever earn the praise and respect of my parents, Ramanujan's story offered a glimmer of hope. It showed me that there might be a way to earn my parents' respect that didn't require following the rigid script that they had written for me—the single-minded pursuit of academic credentials.

I had decided to take my chances and reject my parents' rigid formula for success, just as I had quit the violin cold turkey, and just as I had chosen to race bicycles against their wishes. But I still wanted parental approval. I wanted my parents to accept my decision.

For them, earning a high-school degree is a minor achievement that isn't worth celebrating. After all, the vast majority of kids graduate, so what's the big deal? And so paradoxically, it was my parents' low regard for high school that kept my argument in favor of dropping out alive.

In a last-ditch effort, after months of heated shouting matches, I played the Ramanujan card. I offered him as a role model, a successful dropout whom my

© Springer International Publishing Switzerland 2016
K. Ono, A.D. Aczel, *My Search for Ramanujan*, DOI 10.1007/978-3-319-25568-2_5

parents understood and revered. I had no expectation that they would accept such a flimsy argument after rejecting every other justification I could throw at them, but to my surprise, my father, who almost never expressed a firm opinion, soon agreed that I didn't need to finish high school to get into a top college. Although my mother didn't believe a word of it and remained firmly opposed to my plans, she was once again outranked. And thus it was that my parents, exhausted from my arguments and our earlier battles over the violin and cycling, gave me, if not their blessing, at least their permission to make my escape.

I was astounded that my ploy, offering Ramanujan as an example of a successful dropout, had worked. I couldn't believe it. I had offered someone who was nothing like myself, a man who had lived in a faraway country in a faraway time, and who was producing astounding original research when he dropped out of college. Surely I, who was dropping out of high school having produced nothing, wasn't in the least like him.

What made my father give in? I turned the mystery over and over in my mind. I finally decided that he had simply become exhausted from our endless battles. I supposed that he had relented just to get rid of me and finally have some peace and quiet. But I was wrong, though a dozen years would pass before I discovered the real reason.

A few months later, with my parents resigned to my decision, I moved out. They dropped me off at Penn Station in Baltimore, where I boarded a train bound for Montreal. I had my fancy Peugeot, and some clothes stuffed in an old blue canvas suitcase held together with duct tape.

Who would have thought that some long-dead Indian dude named Ramanujan would be the talisman that would unlock the door of my cell? I had no way of knowing then that I would be hearing a lot more from that miraculous mathematician, that one day I would be drawn to search for the source of his mathematics, a search that would finally let me come to terms with my tough-loving Japanese tiger parents.

Part II
The Legend of Ramanujan

Chapter 6

~

LITTLE LORD

Kumbakonam, India (1890s)

rowds ramble by on foot and rickshaw on this busy dirt road lined
with small shops and peddlers seated on blankets selling vegetables,
silks, tin plates, and almost anything else you might imagine. Most of
the people are thin Indians clad in light, loose-fitting garments. Once in a while,
a British officer or administrator passes by in a khaki uniform or formal suit.
Cows and goats roam freely, eating whatever scraps and waste they can find in
this hot and humid place. This is Sarangapani Street in late-nineteenth-century
Kumbakonam, a town in the southeast Indian state of Tamil Nadu, Ramanujan's
childhood home.

The year is 1887, and Komalatammal, Srinivasa Ramanujan's mother, has left
her husband behind in Kumbakonam to travel the 150 miles west to her mother's home in Erode to give birth to her first child. She is a corpulent, authoritar-
ian woman who, like my mother, runs her household with little assistance from
her husband, who works as a clerk in a fabric store. And so it was in Erode, on
December 22, 1887, that Srinivasa Ramanujan Iyengar was born. A year later,
mother and son returned home to Kumbakonam.

Srinivasa was the baby's father's name, and Iyengar indicated his high priestly
Brahmin caste. His father was K. Srinivasa Iyengar, and so the future mathematician would be known throughout his life simply as Ramanujan (the emphasis is
on the middle syllable, Rah-MAN-ujan), the only part of his name that was
uniquely his. The name derives from the Indian mythological hero Rama, on
whose life and deeds the famous Indian epic the *Ramayana* is based. His doting
mother, however, referred to him throughout his life by the endearing nickname
Chinnaswami, which means "Little Lord." And despite the family's poverty, he
grew up like one.

© Springer International Publishing Switzerland 2016
K. Ono, A.D. Aczel, *My Search for Ramanujan*, DOI 10.1007/978-3-319-25568-2_6

Map of South India (drawn by Aspen Ono)

Like another genius, Albert Einstein, born eight years earlier, Ramanujan began to talk unusually late—to the point that his parents worried whether he would ever be able to communicate. His mother feared that he was deaf, and following a suggestion made by a friend of her father's, he was made to study the characters of the Tamil language—thus learning to read before he ever opened his mouth to speak. Finally, having memorized the characters of this ancient, linguistically pure Dravidian language, Ramanujan began to sound them out, and finally began to talk.

Ramanujan's mother, Komalatammal

Ramanujan's father was rarely home—leaving early in the morning and returning home late in the evening—and like my father, he paid little attention to his son. Ramanujan was raised by his strong-willed mother, who helped him with his schoolwork and even advocated on his behalf by complaining volubly to the school principal whenever she was unhappy with how her son was being treated in school. Ramanujan grew to be a willful child who rarely agreed to do anything that didn't suit his wishes, and he routinely ignored instructions from teachers and school administrators, indeed from anyone in authority. He was also exceptionally sensitive, taking offense where it may not have been intended, and possessing a perfectionistic attitude and a keen sense of shame. For him, any personal failure, however slight, was followed by severe mental anguish. If one of his classmates achieved a higher score on an exam in an area that he believed he understood better, Ramanujan would agonize over his "defeat."

Komalatammal would later give birth to three more children, but all of them died shortly after birth. That, unfortunately, was far from an unusual circumstance. Many children born in south India in that period died at birth or at a young age. Other children would be born to the family many years later, after Ramanujan had reached young adulthood, and so he was raised virtually as an only child.

His mother, like mine, held a part-time job. She worked in a temple choir, singing hymns to the local goddess. This gave the family some much-needed extra income to supplement the father's meager earnings. The family was poor, but richer than the average family in the peasant class of south India. As Brahmins, members of the highest Indian caste, they lived with pride despite their economic difficulties.

Kumbakonam, which had about fifty thousand inhabitants at the time, is located within a vast agricultural region watered by the Cauvery River—a water source considered almost as holy as the famous Ganges hundreds of miles to the north. Mosquitoes proliferated in this area of ample standing water, and so malaria, a scourge that kills the young as well as the old—was rampant. So were many other diseases. A smallpox epidemic raged in this region in 1889, when Ramanujan was two years old, killing thousands of children. The boy fell seriously ill, but he beat the odds and survived.

Just before reaching the age of five, Ramanujan enrolled in school in Kumbakonam. He disliked school. With the personality of a "Little Lord," it was not easy for him to accept authority, and because he was an exceptionally bright child, his mind was logical and critical, and he was particularly unwilling to do anything that was asked of him for which he could see no good reason. Like me, he excelled in mathematics, and during this time in his schooling he was strong in all of his subjects, and he gained skills that would prove useful in later life. One of those skills was proficiency in English. Only about a tenth of the population of India at that time had good command of the English language—which was important for communication among the peoples in the subcontinent, who spoke well over one hundred different major native languages, most of them Dravidian or Indo-Aryan in origin. A proficiency in English opened up for a young person the possibility of obtaining a job in the civil service or some other well-paying profession.

A month before his tenth birthday, Ramanujan took exams in Tamil, English, geography, and arithmetic. He scored first in the entire district and was praised for his outstanding performance. That achievement enabled him to enroll in the local high school, Town High, one of the best schools in the area. Classes were taught in English, which was a great advantage to its students.

Already in the second year of high school, Ramanujan became known for his mastery of mathematics, and students from the school would often come to him for help with problems. In mathematics, he reigned supreme, and his knowledge quickly approached that of his teachers. Through books he had obtained, Ramanujan learned a great deal of trigonometry, geometry, and algebra on his own.

Under the overbearing guidance of his mother, Ramanujan emerged as a headstrong and brilliant ten-year-old. He so excelled in his studies that nobody would have believed that the day would come when school would be a struggle for him.

Chapter 7

~

A CREATIVE GENIUS

*J*ust like the kids I knew from Julian Stanley's John Hopkins study of talented children, Ramanujan was identified at an early age for his gifts in mathematics. Like us, he was capable of flying through the usual mathematics curriculum. However, where we had a knack only for understanding formulas, he had a knack for creating them.

On his own, Ramanujan was able to see past the formulas to the theory behind them. Indeed, he quickly exhausted the mathematical knowledge of his teachers at school and began to read books on mathematics independently, which he obtained from two university students who boarded with his family, who welcomed the additional income, which supplemented what Srinivasa was making as a clerk and what Komalatammal was paid for her temple singing.

As Ramanujan was growing up, Komalatammal was spending less and less time with the choir, feeling that she needed to focus on nurturing her young lord. More than anyone else, she understood how brilliant her son was, and she would spend hours teaching him about the world, educating him at home in ways that were more suited to his personality than the rigid rote instruction that he was receiving at school. She played intellectual games with him, took him on long walks by the river, and even consulted with him about her own life. They were exceptionally close.

Ramanujan quickly devoured the books lent to him by the two boarders, teaching himself so much that he had now advanced well beyond what the teachers at his school could offer. And he began to go beyond the mathematics presented in the books he had been given, developing his own ideas about trigonometry, geometry, number theory, and infinite series.

In 1900, while only in the second year of secondary school (corresponding to our seventh grade), Ramanujan began to figure out on his own how to work with infinite series, that is, how to efficiently represent the sum of infinitely many

© Springer International Publishing Switzerland 2016
K. Ono, A.D. Aczel, *My Search for Ramanujan*, DOI 10.1007/978-3-319-25568-2_7

terms, which may or may not represent a unique finite number. Infinite series and continued fractions would become an obsessive occupation for him throughout his life—as they would, decades later, in my own work.

While still a child, Ramanujan was fast becoming an expert on those seemingly intractable mathematical entities. He spent more and more of his time trying to understand how infinite series work, which ones converge and which ones do not. He would soon find a method that allowed him to add up all of the positive integers, that is, the expression $1+2+3+4+5+6+\cdots$, and obtain a *finite negative* number! As crazy as this claim appears to be, one can make complete sense out of it, and it requires important theorems by some of the world's great mathematicians, including Jacob Bernoulli, Leonhard Euler, and Bernhard Riemann, on a subject called the "analytic continuation of the Riemann zeta function." The young, untrained Ramanujan had somehow obtained glimpses of the work of Bernoulli, Euler, and Riemann on his own.

Ramanujan had such an amazing aptitude for mathematics that by age twelve, he had worked out new solutions to problems in number theory and analysis. Astonishingly, he was able to come up with mathematical facts and ideas in what was close to an intellectual vacuum.

India had a long tradition, going back to the early Middle Ages, of producing important mathematical results without proof. And like some other Indian mathematicians, Ramanujan cared little about formal proof. He simply derived beautiful mathematics as if out of thin air—mostly identities and equations.

In 1902, he learned about the method that Italian mathematicians had discovered in the sixteenth century for solving the cubic equation, and he derived on his own a method for the solution of quartic, or fourth-degree, equations—repeating a variation of a feat performed brilliantly by René Descartes three centuries earlier (though Descartes was much older at the time than Ramanujan).

Ramanujan managed not to antagonize his teachers too much while still in school. But his mathematical prowess was growing without bound and consuming more and more of his time. Like Descartes, Galois, and Einstein, Ramanujan became known in his school as a math genius. He kept winning awards for excellence, and in 1904, he won the K. Ranganatha Rao Prize in mathematics. In announcing the honor to the gathered teachers and students, the principal, K. Iyer, described Ramanujan's abilities in mathematics as being above an A+ level. At his graduation from Town High, Ramanujan won a scholarship to attend Government College, an excellent institute of higher education in Kumbakonam, considered by some the "Cambridge" of the region. At

Government College, Ramanujan continued to perform amazingly well in mathematics, but he now began to do poorly in everything else. There was an unusual reason why this brilliant student, who was far better than everyone else around him in mathematics and whose knowledge exceeded even that of his teachers, stopped showing even minimal interest in any other subject.

Chapter 8

⌒

AN ADDICTION

*I*magine a college student who is so consumed by video games that he is failing all his classes. He even forgets to eat, shower, and sleep. If we replaced the video games by equations and formulas, then this would have been Ramanujan.

During the last few months of his time at Town High School, Ramanujan became obsessed with mathematics. So much so, in fact, that had the obsession taken hold of him earlier, he might have jeopardized his graduation, because his addiction left room for nothing else in his life, no other subject whatsoever. What triggered his strange state of mind was a book published in 1886 titled *A Synopsis of Elementary Results in Pure Mathematics,* by George Shoobridge Carr. Carr was a professional mathematics tutor in London, and he had decided to put in book form all the formulas, theorems, and results known during his time that would be useful in tutoring English mathematics students.

A friend had borrowed a library copy of this book for Ramanujan, and it is this book that appears to have unleashed Ramanujan's true genius, and it has thus played an unlikely role in the history of mathematics. Carr was by no means a great mathematician, and his book—a compendium of results known in his time—was later described by the mathematician Godfrey Harold Hardy (who will play a major role in this story) in a 1937 article on Ramanujan as follows:

> *The book is not in any sense a great one, but Ramanujan has made it famous, and there is no doubt that it influenced him profoundly and that his acquaintance with it marked the real starting point of his career.*

© Springer International Publishing Switzerland 2016
K. Ono, A.D. Aczel, *My Search for Ramanujan,* DOI 10.1007/978-3-319-25568-2_8

Carr's book contained 6165 theorems and mathematical formulas, beginning with relatively simple ones, such as $a^2 - b^2 = (a+b)(a-b)$, and continuing to increasingly more complicated results. The characteristic trait of this book—unlike others that Ramanujan had read while in school—was its extreme terseness. The *Synopsis* contained few details of proofs. This had the distinct effect of encouraging Ramanujan to find "proofs" himself, and thus to develop his own mathematical chops. Within months—at which time he was a scholarship student at a prestigious local college—Ramanujan had become hooked on Carr and could focus on nothing else.

While Carr's highly condensed book allowed Ramanujan's genius to emerge because he felt the need to supply the details in the theorems, and thus to explain to himself the reasoning behind the results Carr had presented, it had several negative effects. First, Carr had begun tutoring in the 1860s, and what he was preparing his pupils for was the Tripos examination, which required a great deal of rote learning of basic material that university students needed to know to obtain a bachelor's degree. None of this was cutting-edge mathematics. Thus Ramanujan's introduction to "modern mathematics" was not at all modern. It may have given him a wrong impression about the state of the art—like my father in Japan at the end of World War II, he was isolated from current trends—and when, some years later, he would greatly extend the formulas in Carr's book, thinking perhaps that he was doing highly original work, it would appear to others that he was simply appropriating known facts and passing them off as his own.

Second, the book's extreme terseness—it was, after all, primarily a collection of mathematical facts that an undergraduate would need to know—may have given Ramanujan the false impression that this was how mathematics was done in the wider world, namely that there was no need of formal argumentation. Thus, when Ramanujan began to create his own mathematics, he, too, would provide no justifications for his results in the form of detailed derivations, as had become custom in mathematical writing since the mid-nineteenth century.

The third problem with the book was perhaps the most unfortunate of all. It was too absorbing, too fascinating, and it deprived Ramanujan of the chance for a solid education. He discovered the book in late 1903, and he became totally obsessed with it. He spent all his time trying to supply the mathematical details left out by Carr.

Ramanujan managed to coast during his last few months of high school without slipping too badly in any subject, even though he was spending all his time on his mathematical derivations and extensions of Carr's material. He received his awards and a scholarship, which was badly needed, since his parents' income was much too low to afford him a college education.

Ramanujan entered Government College in 1904. From the outset, his situation was not auspicious. He continued to do nothing but mathematics. Nothing else interested him. He would sit in class, pretending to listen to the lecturer, with Carr's *Synopsis* in his lap, his mind deep in thought about an infinite series or an infinite continued fraction. To what did these series or fractions *converge*? he would ask himself, completely unaware whether the professor was talking about the history of India or a Shakespearean tragedy. There was a curriculum that he was supposed to be following, but Carr's book, and the flood of mathematical ideas that it inspired, had eclipsed everything else.

Ramanujan was just a teenager, but he was already developing original research based on what he was reading. Of course, he was working in a relative mathematical desert, for despite its prestige in southeast India, Government College, with its mere dozen lecturers, was an intellectual backwater. Although unaware of what was known or unknown beyond the pages of Carr's book, and although some of Ramanujan's results, while original to him, were well known to the mathematical world, some of his results were new, often going beyond those obtained by some of the world's great mathematicians, such as the brilliant eighteenth-century Swiss mathematician Leonhard Euler, arguably the most prolific mathematician of all time, whose collected mathematical papers fill about ninety volumes.

So although Ramanujan dutifully attended the required lectures, his mind was elsewhere—engrossed in the riddles of Carr's book, and "quite unmindful of what was going on around him," as a classmate later described him to one of Ramanujan's biographers. Although in high school he had excelled in many subjects, including English, he now failed his English composition paper, and as a result, he lost his scholarship. His mother went to the head of the college to plead, beg, cajole, and complain. But the scholarship was not reinstated.

Ramanujan remained in school a few months longer, but without the scholarship, college was unaffordable. Torn between family loyalty, love of mathematics, the desire to please, and the pull of a greater force, Ramanujan was lost. He did the only thing he could, something with which I readily sympathize: he ran away from home. He took a train to the distant town of Vizagapatnam, seven hundred miles up the coast, leaving without a word at the beginning of August 1905. His family made frantic inquiries. They advertised and posted notices for their missing son, and they soon found him and brought him home.

Ramanujan attempted to return to university to continue his education. In 1906, he traveled the two hundred miles by train to Madras, arriving dazed and confused, but eventually making his way to Pachaiyappa's College, where he hoped to study and obtain his degree. Here, the experience from Kumbakonam repeated itself: Ramanujan excelled beyond expectations in mathematics—easily surpassing the knowledge of his professors, who held him in awe. But he just couldn't concentrate on his other subjects. He failed his physiology exam miserably several times, and eventually had to withdraw from the college without a degree.

Chapter 9

∼

THE GODDESS

*M*athematically talented children are frequently identified as such by their ability to perform very large calculations rapidly. Ramanujan was a prodigious calculator, but what set him apart was his creativity—his ability to conjure never before imagined mathematical formulas out of thin air. Where did they come from?

Despite his status as an untrained amateur mathematician, and a college dropout to boot, the young Srinivasa Ramanujan was producing mathematical results that were so unexpected and so significant that they are striking to a mathematician who sees them for the first time even today, a century later. When I first saw them, they appeared to me as exquisite mathematical treasures, which had been revealed to Ramanujan as if by magic, as though spirited from the depths of some Ali Baba's cave. And even at an early stage of my life as a professional mathematician, I found them utterly irresistible, and I would eventually tie my career to them. I just had to find out how Ramanujan had obtained them.

How could he possibly know that the mathematical formulas he was conjuring were correct —formulas that would take some of the world's leading mathematicians months and years to prove? There was something mystical, supernatural, perhaps even spiritual in the way Ramanujan obtained his results, and indeed, those are the sorts of adjectives that are frequently applied to him.

When asked how he obtained his results, Ramanujan would reply that his family goddess, Namagiri, sent him visions in which mathematical formulas would unfold before his eyes.

Brahmins were at the top of India's caste system, and they prided themselves on their historical role in perpetuating and maintaining the Hindu religion. Ramanujan was a devout Hindu—likely not only because he was a Brahmin, but also through the influence of his mother, who was deeply involved with activities

© Springer International Publishing Switzerland 2016
K. Ono, A.D. Aczel, *My Search for Ramanujan*, DOI 10.1007/978-3-319-25568-2_9

in the local temple to the goddess Namagiri. The Sarangapani temple was just up the street from Ramanujan's family home in Kumbakonam, and he spent much of his time there doing his mathematics.

He once explained to a friend that he saw in the mathematical expression $2^n - 1$ "the primordial God and several divinities." This was because when one plugs $n = 0$ into the expression, one gets $1 - 1 = 0$, which represents nothingness. Plugging in 1 for n gives the result $2 - 1 = 1$, which he said represented to him unity and the infinite God. When substituting 2 for n, one obtains 3, the Trinity, which to him meant the Hindu gods Shiva, Vishnu, and Brahma. Plugging in 3 for n yields the number 7, which represented the Saptha Rishis—the "Seven Sages" of Hinduism, described in the Vedas, sacred texts dating to as early as the second millennium B.C.E. Ramanujan also found religious symbolism for additional values of n.

He once said to a friend, "An equation for me has no meaning unless it expresses a thought of God." In this, he came close to an almost identical statement by Einstein, who in describing his work on general relativity and his attempts to find a theory that would capture the nature of gravitation and the physical laws of the universe, famously said, "I want to know God's thoughts."

Beginning in 1907, having lost his college scholarship and at loose ends regarding everything in his life outside the pages of Carr's book, Ramanujan began to keep a notebook of his mathematical results. He claimed that they came to him in dreams and visions: Namagiri would appear to him and draw mathematical statements over a red screen. In the morning, he would write them down, ponder them, analyze them, and study them further to see how they could be extended to new results. In fact, he would often wake up in the middle of the night to write down a formula that had appeared to him in a dream, and in the morning, he would write it out in greater detail.

When I first heard this story about Ramanujan and his dreams, I rejected at once as fanciful poppycock the suggestion that his insights came to him as visions from a goddess. I, who had been raised with no religion and a strong belief that all phenomena have a rational explanation, viewed the idea of divine inspiration as a fable invented by others to add mystery to the legend of Ramanujan, or else cooked up by Ramanujan himself to elevate his accomplishments beyond the mundane world of hard work, the sort of thing my father did, scribbling all day on yellow pads of paper.

But I have gone a very long distance out of my way since then, and today, I have quite different views.

Chapter 10

~

PURGATORY

*G*enius often finds itself rejected when it fails to fit into the mold designed for ordinary people. Einstein, for example, was at first denied admission to the prestigious Swiss Federal Institute of Technology, in Zurich, because he failed the nonscience portion of the entrance examination. Thomas Edison's teachers told him that he was too stupid to learn anything, and he was fired from his first two jobs for not being productive enough. Ramanujan was in good company.

When we last saw Ramanujan, he had just flunked out of Pachaiyappa's College. He had hoped eventually to enter the University of Madras, since a degree from that university would have qualified him as a working mathematician or given him the credentials to obtain some other respectable job with a decent salary. But having failed the physiology part of his exams repeatedly, he now found himself out of school and unemployed. No work meant no money, and Ramanujan did not have even enough to feed himself. In desperation, he returned home to Kumbakonam. But soon he was on the road again.

Returning to Madras, Ramanujan continued to live in abject poverty. He had little to eat and was desperate to find some kind of employment. He began looking for any sort of job, and he finally obtained a temporary post as a clerk, but he left after a few weeks. He couldn't keep his mind on his work; his addiction to mathematics was too strong. Then he decided to try to obtain a more suitable position—one that he hoped would leave him time for mathematics—by contacting men in the British administration of India, the famously efficient Indian Civil Service, who were also mathematicians or had some knowledge and appreciation of the subject. He began to build a network, contacting friends, friends of friends, family, family friends, and anyone who knew anyone who knew anyone who might appreciate his mathematical genius and decide that it would be a shame for someone so talented to be without a livelihood.

© Springer International Publishing Switzerland 2016
K. Ono, A.D. Aczel, *My Search for Ramanujan*, DOI 10.1007/978-3-319-25568-2_10

Ramanujan had a pleasant personality. He was uncommonly likeable and engaging, which made even those who had met him only briefly happy to give him a reference. Given the name of someone who might be able to help him, he would go and knock on his door. He also approached professors at the colleges he had attended as well as at other colleges and universities in south India, showing them his mathematical work and hoping to interest them in providing him a post that would enable him to continue to do mathematics.

His work was so astonishingly novel—but also seriously lacking in detail and anything resembling formal proof—that some of the professors at first doubted that it was his own. But once he showed them how he derived his equations, a few understood that he was not a fraud, and they wrote enthusiastic letters of reference on his behalf—but they came to nothing.

Eventually, as his money dried up, his situation became more dire, and he had to think seriously about nonmathematical subjects, such as where his next meal was coming from. Desperate, he began to look for almost any form of employment. Eventually, he obtained a temporary post as clerk in the Office of the General Accountant, in Madras, a very minor civil service position with a very small salary. He could barely survive, and he continued looking for something better that would allow him time to pursue mathematics without starving for it. Finally, his networking began to pay off, and he went to meet a truly important man, Ramachandra Rao, who held a high position as revenue collector in the city of Nellore. And what was more, he was a mathematician, and indeed was currently serving as secretary of the Indian Mathematical Society. In a 1920 article in the *Journal of the Indian Mathematical Society*, Rao described his first meeting with Ramanujan:

> *Several years ago, a nephew of mine perfectly innocent of mathematical knowledge said to me, "Uncle, I have a visitor who talks of mathematics; I do not understand him; can you see if there is anything in his talk?" And in the plenitude of my mathematical wisdom, I condescended to permit Ramanujan to walk into my presence. A short uncouth figure, stout, unshaved, not overclean, with one conspicuous feature—shining eyes— walked in with a frayed notebook under his arm. He was miserably poor. He had run away from Kumbakonam to get leisure in Madras to pursue his studies. He never craved for any distinction. He wanted leisure; in other words, that simple food should be provided for him without exertion on his part and that he should be allowed to dream on.*

He opened his book and began to explain some of his discoveries. I saw quite at once that there was something out of the way; but my knowledge did not permit me to judge whether he talked sense or nonsense. Suspending judgment, I asked him to come over again, and he did. And then he had gauged my ignorance and showed me some of his simpler results. These transcended existing books and I had no doubt that he was a remarkable man. Then, step by step, he led me to elliptic integrals and hypergeometric series and at last his theory of divergent series not yet announced to the world converted me. I asked him what he wanted. He said he wanted a pittance to live on so that he might pursue his researches.

Rao was taken enough with the brilliance of Ramanujan that he determined that offering him a position in Nellore would not do—Ramanujan belonged in the larger Madras, where he could meet mathematicians and others—and instead offered him a stipend that would allow him to continue his mathematical work unhindered by worries of employment. This was a godsend. Ramanujan settled in Madras and continued to work on his identities.

At some point, he returned home to his parents in Kumbakonam. There, completely oblivious to what was going on around him, he would sit on the porch, doing mathematics in total concentration, his eyes shining when he made a discovery. His parents supported him without complaint and showed no irritation or inclination to push him to look for a job. But his mother had certain ideas of her own about his future.

Chapter 11

~

JANAKI

What Ramanujan's mother had in mind was a wife for her son, and like my parents' marriage, Ramanujan's was to be an arranged one. After consulting the goddess Namagiri, Komalatammal had decided that it was time for her son to get married. So she set out to find him a wife. Through an Indian tradition allowing child marriages, whose consummation was deferred, Komalatammal found her son a child bride, the ten-year-old S. Janaki Ammal.

His mother had arranged everything. She began with a visit to distant relatives in Rajendram, seventy miles west of Kumbakonam, a village so small that it does not appear on most maps. There, she noticed a sprightly girl of nine whose family had five girls to marry off and was so poor that they could offer little by way of a dowry. She immediately consulted the girl's horoscope in conjunction with that of her son—as was the custom in such matters—and decided that Janaki was perfect for Ramanujan. A few months later, on July 14, 1909, when Janaki was ten, the wedding took place in Rajendram, and at the same ceremony, one of Janaki's sisters was also married. Unfortunately, this sister would die of a fever some months later.

After the wedding, Ramanujan returned home with his mother to Kumbakonam, and over the next few years he would become a mathematical nomad—traveling by train throughout south India, to Madras, and elsewhere—while Janaki stayed home with her mother, who was teaching her how to be an obedient wife and perform household chores. The couple would not live together until 1912, in Madras, where Komalatammal would live with them.

It was this child bride of Srinivasa Ramanujan who, seventy-five years later, as an eighty-five-year-old widow, sent my father that fateful thank-you letter.

© Springer International Publishing Switzerland 2016
K. Ono, A.D. Aczel, *My Search for Ramanujan*, DOI 10.1007/978-3-319-25568-2_11

Janaki Ammal

Chapter 12

I BEG TO INTRODUCE MYSELF

*T*hough he was without a college degree and had essentially no formal training in mathematics, Ramanujan had accumulated a massive collection of formulas, all recorded in his notebooks without proof. Eager to share his work with others, he began to publish some of his findings, beginning in 1911 by submitting problems to the *Journal of the Indian Mathematical Society*. At first, nobody paid attention, or at least no one was able to solve his problems. For example, one of the problems that he challenged readers to solve was to find the value of

$$\sqrt{1+2\sqrt{1+3\sqrt{1+\cdots}}}.$$

What number was represented by this infinite nested square root?

Six months passed, and three new issues of the journal appeared, yet no one had proposed a solution. If any readers were paying attention, they were stumped.

So Ramanujan simply revealed the answer: 3. But he did not reveal his methods, and in fact, finding the solution is difficult, requiring a long derivation. Ramanujan had developed a formula that took a sum of three numbers and expanded it into an infinite nested square root. The problem he posed was a special case of that formula for the sum $2 + 1 + 0$.

In 1912, Ramanujan finally obtained a more permanent position, as a clerk in the Madras Port Trust. He performed his job so efficiently that he had time to continue his research in mathematics and publish more papers. His brilliance was now more widely appreciated, and Ramanujan and his supporters began to realize that he would never receive the recognition he deserved in India, where little cutting-edge research was being done. Most Indian mathematicians were

© Springer International Publishing Switzerland 2016
K. Ono, A.D. Aczel, *My Search for Ramanujan*, DOI 10.1007/978-3-319-25568-2_12

either pursuing mathematics as an avocation on top of another job or working in a college or university with heavy teaching duties and little time or reward for research. They would not have the background to understand and appreciate Ramanujan's mathematical accomplishments.

That same year, 1912, Janaki finally came to Madras to live with the man who was formally her husband. He had a position now, and could support her, and she was thirteen, on the cusp of adolescence. Komalatammal came from Kumbakonam to live with the couple, further training Janaki in the occupation of housewife. She was a constant chaperone, and when she had to be absent, another woman would take her place in the household.

Ramanujan had obtained his post at the Port Trust through a recommendation by Ramachandra Rao, his staunch patron in Nellore. His supervisors in Madras were the Englishman Sir Francis Spring and his subordinate, the Indian S. Narayana Iyer. His work, which consisted in reviewing accounts of the trust, allowed him time to pursue his mathematical research, and both his supervisors encouraged him in that direction. For the first time in his life, Ramanujan had satisfactory employment, was appreciated both for his performance at work and his mathematical endeavors, and could devote time to filling his curious notebooks with a treasure of mathematical identities. But both Sir Francis and Narayana Iyer understood that Ramanujan needed an audience for his work outside of India, since no Indian mathematician could fully understand and appreciate his work, and as an Indian, Ramanujan would unlikely receive recognition by the British establishment in India, which maintained a strict color bar. They decided to appeal to mathematicians in Britain on Ramanujan's behalf.

They began by writing to Micaiah John Muller Hill, Astor Professor of Mathematics at the University of London, who had served as chancellor of the university and was a Fellow of the Royal Society. Hill wrote back courteously, suggesting that Ramanujan be more careful with his notation and the presentation of his results, but he had clearly failed to appreciate Ramanujan's genius, which was buried in his nonstandard notation and paucity of proofs. Disappointed, Ramanujan himself wrote to two British mathematicians, H.F. Baker, a Fellow of the Royal Society and former president of the London Mathematical Society, and E.W. Hobson, also a Fellow of the Royal Society and chair of pure mathematics at Cambridge University. Both returned Ramanujan's submissions without comment.

Ramanujan did not give up. On January 16, 1913, he wrote to G.H. Hardy, the Sadleirian Professor of Mathematics at Cambridge University and also a Fellow of the Royal Society. Unlike Ramanujan, who loved only mathematics, Hardy was interested in many subjects. He had almost decided to study history on entering university; he was interested in music and art; and throughout his life, he was fanatically devoted to the game of cricket

Hardy viewed himself as a "pure mathematician" and boasted that his work had no application to the real world. (Little did he know that one of his main interests, the theory of prime numbers, would become the basis of the encryption algorithms that have become omnipresent in the digital age.) In his 1940 essay *A Mathematician's Apology*, Hardy expounded his views about the importance of the purely intellectual pleasures of mathematics, stripped of any direct application to the real world. He compared proofs of theorems to poems, works of art, and musical compositions.

Hardy was the leading number theorist in the world at that time, and his textbook on number theory, *An Introduction to the Theory of Numbers*, written with E.M. Wright, would become a classic in the field. Moreover, in 1911, Hardy had begun a long and strikingly fruitful mathematical collaboration with his colleague John E. Littlewood, a titan in the field of mathematical analysis. This was a pair eminently qualified to judge Ramanujan's work, and in writing to Hardy, Ramanujan had aimed his dart well.

G. H. Hardy

CHAPTER 12

Madras Port Trust Office
Accounts Department
16th January 1913

Dear Sir,

I beg to introduce myself to you as a clerk in the Accounts Department of the Port Trust Office at Madras on a salary of only £20 per annum. I am now about 23 years of age. I have had no University education but I have undergone the ordinary school course. After leaving school I have been employing the spare time at my disposal to work at Mathematics. I have not trodden through the conventional regular course which is followed in a University course, but I am striking out a new path for myself. I have made a special investigation of divergent series in general and the results I get are termed by the local mathematicians as "startling."

There followed pages of formulas and assertions, all offered without a hint of proof. Ramanujan touched on divergent series, exotic integrals, formulas involving π, hypergeometric functions, prime numbers, divisor functions, sums of squares, and continued fractions. He considered the series of positive integers, namely $1 + 2 + 3 + 4 + \cdots$, and he claimed that it could be interpreted as summing to $-1/12$ instead of infinity, hinting at his rediscovery of results of Bernoulli, Euler, and Riemann. He offered remarkable-looking formulas involving π and the trigonometric functions sine and cosine. Some of his formulas were incorrect, but the errors could be corrected once the idea was understood. Some formulas remained unproved for a long time, while others await a solution by future mathematicians. Some of Ramanujan's expressions are so abbreviated or cryptic that mathematicians today continue to struggle over them.

Ramanujan concluded his letter with a heartfelt request:

Being poor, if you are convinced that there is anything of value I would like to have my theorems published. I have not given the actual investigations nor the expressions that I get but I have indicated the lines on which I proceed. Being inexperienced I would very highly value any advice you give me. Requesting to be excused for the trouble I give you.

I remain, Dear Sir
yours very truly
S. Ramanujan

Chapter 13

⌇

THESE FORMULAS DEFEATED
ME COMPLETELY

*A*nd so it was that toward the end of January 1913, or perhaps in the first days of February, Hardy, over breakfast in his rooms at Trinity College, in Cambridge, received a curious letter from India containing nine pages of Ramanujan's original mathematical work.

A mathematician of Hardy's rank and reputation often received letters from amateurs who thought they had discovered something new in mathematics and were seeking validation, perhaps publication, perhaps even renown. As an editor of several mathematical journals, I too am accustomed to receiving manuscripts from amateurs. Each year, I receive dozens of flawed proofs of famous problems, such as Fermat's last theorem and the Riemann hypothesis, that have stumped generations of mathematicians. (Fermat's last theorem, having resisted solution for almost four hundred years, was finally proved by Andrew Wiles in 1995 using almost the entire armamentarium of advanced twentieth-century mathematics, while the Riemann hypothesis continues to defy the world's greatest mathematical minds.)

Most of these invalid proof attempts fail because the amateur has made a false assumption that if true would render the problem trivial. It's rather like getting trapped in a fool's mate in chess: you think you have mounted a powerful attack against your opponent's king, but the next thing you know, you have been checkmated. Other proofs sent in by amateurs are correct, but they are merely redis-coveries of long-known results.

What Hardy had before him that morning looked like real mathematics. But that didn't mean it was necessarily the real deal. Was it fraudulent? he asked himself. Had this unknown Indian with the careful schoolboy handwriting that always missed crossing his t's perhaps copied the work of a mathematician from an obscure journal and was trying to pass it off as his own? Or was the work perhaps

© Springer International Publishing Switzerland 2016
K. Ono, A.D. Aczel, *My Search for Ramanujan*, DOI 10.1007/978-3-319-25568-2_13

genuinely his own but of no real value? Looking at the formulas more carefully, Hardy recognized some of the results as mathematical derivations that had been obtained by others and were well known. Others made no sense to him at all. Yet others were so fantastic that they had to be the work of either a crank or a genius. It would just have to wait. He put the letter aside, picked up his newspaper, and continued with his breakfast, planning to look at it more carefully that evening.

After the morning's regimen of four hours of mathematical research came lunch, then perhaps a game of tennis and dressing for dinner. That evening, Hardy again studied the letter from India. He became more and more intrigued. As he reread the pages, he realized that two results that had at first looked like nonsense were somehow related to hypergeometric series and continued fractions, which had been previously studied by Euler and Gauss.

Continued fractions are numbers that are described by an iterative process of division. For example, we can express the fraction 7/10 as a continued fraction thus:

$$\frac{7}{10} = \frac{1}{\frac{10}{7}} = \frac{1}{1+\frac{3}{7}} = \frac{1}{1+\frac{1}{\frac{7}{3}}} = \frac{1}{1+\frac{1}{2+\frac{1}{3}}}.$$

Such a finite continued fraction always represents a rational number. Things get interesting when irrational numbers are developed into infinite continued fractions. Let us take as an example a famous irrational number, the *golden ratio*, denoted by the Greek letter phi,

$$\phi = \frac{1+\sqrt{5}}{2},$$

though any old square root would have done almost as well. Pythagoras, Euclid, and Kepler, among many other distinguished scientists through the ages, have been fascinated by this number. It is called *golden* because as a proportion, it is pleasing to the eye and has been used extensively in works of art. It also appears in nature, in such diverse places as the arrangement of leaves and branches of plants, the geometry of crystals, and the structure of DNA.

The golden ratio also appears in many of mankind's most beautiful creations. Some great works of architecture and art, such as the Parthenon, the pyramids of Giza, and the *Mona Lisa*, make use of the pleasing proportions of the golden ratio. Salvador Dalí, the famous twentieth-century Spanish surrealist, explicitly referenced the golden ratio in his masterpiece *The Sacrament of the Last Supper*, which depicts Jesus and his disciples seated below a dodecahedron, a geometric figure whose proportions define the golden ratio.

Salvador Dalí's *The Sacrament of the Last Supper* (courtesy of the National Gallery of Art, Washington, D.C.)

As a purely mathematical object, the golden ratio has many faces. It is, for example, the limit of the ratios of successive terms of the famous Fibonacci sequence: 1, 1, 2, 3, 5, 8, 13, 21, 34, 55, 89, ... (each term is the sum of the previous two terms), namely the ratios 1/1, 2/1, 3/2, 5/3, 8/5, 13/8, 21/13, 34/21, 55/34, 89/55, ... The golden ratio was described by Euclid around 300 B.C.E. by "extreme and mean ratios," which leads to the relationship

$$1+\frac{1}{\phi}=\phi, \quad \text{or} \quad \phi^2-\phi-1=0,$$

whose positive solution is

$$\phi=\frac{1+\sqrt{5}}{2}.$$

Despite these elegant descriptions, the infinite decimal expansion

$$\phi=1.6180339887498948482045868343 65\ldots$$

exhibits no apparent pattern. However, with some mathematical sleight of hand, a beautiful pattern can be made to emerge. Using repeatedly the relation

$$1+\frac{1}{\phi}=\phi,$$

we can develop ϕ into an infinite continued fraction:

$$\phi=\cfrac{1}{1+\phi}=\cfrac{1}{1+\cfrac{1}{1+\phi}}=\cfrac{1}{1+\cfrac{1}{1+\cfrac{1}{1+\phi}}}=\cfrac{1}{1+\cfrac{1}{1+\cfrac{1}{1+\cfrac{1}{1+\ddots}}}},$$

and so on ad infinitum. In his first letter to Hardy, Ramanujan indicated that he had found a theory that generalizes this well-known classical fact about the golden ratio. He claimed that he knew how to evaluate a much more general continued fraction, one containing a variable quantity that in theory can represent infinitely many continued fractions at once. Such continued fractions were well known to mathematicians, having been discovered in 1896 by the British mathematician Leonard Rogers. Thanks to the variable, the Rogers–Ramanujan continued fraction, as it is known today, generalizes the golden ratio, which is the special case in which the variable assumes the value 1.

Ramanujan offered two examples, with the variable equal to $-e^{-\pi}$ and $e^{-2\pi}$, where $e = 2.718281\ldots$ is Euler's number. He claimed that the two resulting continued fractions are in fact equal to simple expressions involving the golden ratio. Immediately following these two formulas in the letter, Ramanujan dropped a bombshell. He made the astonishing claim that he had a general method for computing these continued fractions *explicitly* for a large class of values of the variable. If that were true, then Ramanujan possessed a theory unlike anything anyone had seen before.

Hardy was amazed. Of all the values to choose from, why did Ramanujan choose the numbers $-e^{-\pi}$ and $e^{-2\pi}$? How did he then figure out their corresponding continued fractions? Hardy didn't have a clue, prompting him later to remark,

They defeated me completely, I had never seen anything in the least like this before … A single look at them is enough to show they could only be written down by a mathematician of the highest class. They must be true because no one would have the imagination to invent them.

Hardy was not alone in his admiration. These identities have fascinated scores of mathematicians, serving as the topic of hundreds of research papers over the past century. The problem of discovering their source, a general theory that explains them as part of an infinite framework of identities, would become one of my main goals as a mathematician.

Was this a startling new result or a bunch of nonsense? Hardy decided that he had better show the letter to his colleague Littlewood. Littlewood concurred that this curious letter seemed to contain some quite amazing results. The two mathematicians pored over the theorems and formulas in the letter for some time. They were captivated. They met a few more times, trying to make sense of the theorems and series, much of it written in nonstandard notation.

As the days passed, and Hardy and Littlewood, and others to whom Hardy showed the letter, analyzed its contents, Hardy came to the conclusion that they were in the presence of something most unusual. First of all, the letter contained some mathematical results that he and Littlewood had seen before—albeit written in what seemed like a different language, because of the notation—but which, they decided, Ramanujan had likely not seen elsewhere. He had apparently rediscovered published results that had seen little circulation.

Other results were completely new, and they were able to prove some of them, though it took considerable effort. Some assertions were incorrect, but such mistakes were rare, and even the erroneous results were clever, obtained through an interesting mathematical process. It appeared that Ramanujan had developed his own unique methods.

It was clear that Ramanujan had discovered mathematical gems beyond the imagination, and Hardy was not going to let those treasures go to waste. But first, he wanted to see how Ramanujan had justified his results, for Hardy was a strong advocate of mathematical rigor. On February 8, he wrote to Ramanujan expressing his interest, but asking him kindly to supply proofs of his statements:

> *I was exceedingly interested by your letter and by the theorems ... You will however understand that before I can judge properly of the value of what you have done, it is essential that I should see proofs of some of your assertions.*

Chapter 14

~

PERMISSION FROM THE GODDESS

Ramanujan was elated finally to receive a positive response from England, and he promptly replied to Hardy: "I have found a friend in you, who views my labours sympathetically." But instead of proofs of the sometimes audacious statements he was making about infinite series, continued fractions, and integrals, Ramanujan sent Hardy *more* theorems. Hardy received these with disbelief. How could one young man, in a country so distant from the world centers of mathematics, come up with so many ingenious mathematical results? It was a mystery. But he persisted in requesting proofs. And Ramanujan persisted in not supplying them—instead sending even more results.

Hardy began to develop a suspicion. Perhaps the Indian was worried that the English mathematician—whom he knew not at all—would steal his work and publish it as his own. And so Hardy wrote again, explaining that Ramanujan now possessed letters from him, which—if Hardy tried to pass his work off as his own—could be used to incriminate him in a clear case of academic theft. Perhaps this suspicion about a suspicion caused a cooling off of the epistolary relationship that had developed between them. For a while, there were no more letters.

Hardy had to find out where these claims and formulas—marvelous, mysterious, beguiling—came from. Were they true, these mathematical treasures? And if so, why were they true? And how had this stranger on the other side of the world obtained them? Ramanujan would have to come to Cambridge.

One of Hardy's youngest colleagues at Trinity was Eric H. Neville, a man two years younger than Ramanujan. Neville was on his way to India for a research and lecturing project in his area of differential geometry. Hardy asked to speak with him before he embarked for India. Would he be willing, while in India, to travel to Madras, and through academic connections there, try to meet this mysterious young man, a certain Ramanujan? Then, Hardy suggested, if the meeting was satisfactory, he should be convinced to travel to England.

© Springer International Publishing Switzerland 2016
K. Ono, A.D. Aczel, *My Search for Ramanujan*, DOI 10.1007/978-3-319-25568-2_14

Neville arrived in India, and after some time made it to Madras, where he met Ramanujan. The two men—perhaps because they were so close in age—got on very well. In fact, they would become lifelong friends. While in Madras, Neville began gently to coax the reluctant Ramanujan to go to England, where his gifts could be further developed, where he would be appreciated as a mathematician, and where he could work more freely without worrying about a livelihood. Neville reported to Hardy about his modest progress in convincing Ramanujan to think about coming to Cambridge, and Hardy then wrote an official letter of invitation.

Hardy also interceded on Ramanujan's behalf to convince the Indian authorities to support such a move. He described Ramanujan's exceptional abilities in letters to Indian officials at the University of Madras, whose help he needed to secure funding to match what he could garner in England, so that Ramanujan's travel and stay in Cambridge could become a reality. The word spread like wildfire throughout the academic community of south India: a professor of the highest rank in Cambridge, a Fellow of the Royal Society, believed so strongly in Ramanujan that he wanted to bring him to England.

Ramanujan had no such interest. He wrote to Hardy, "What I want at this stage is for eminent professors like you to recognize that there is some worth in me." And now he clearly had obtained the recognition he so badly craved—and everyone in Madras knew it.

Soon, other offers came his way, offers that were much closer to home. Everyone wanted him, everyone hoped that he would stay in his native India and help make it great in mathematics. On April 12, 1913, Ramanujan was granted a generous scholarship as a research student at Presidency College, in Madras. There were no obligations, just that he should pursue his researches in any way he wanted, that he should do the mathematics that he loved with absolutely no other responsibilities. This was truly a dream come true. Hardy's interest in his work had finally won him not only recognition, but the financial independence that would allow him to do mathematics and would grant him also the possibility of receiving—finally, after twice failing to do so—an academic degree that could secure his future. Ramanujan was on top of the world. He had no reason to go anywhere.

When all of the arrangements for his fellowship at Presidency College had been made, he was joined by his mother and his wife, and they found a pleasant apartment near the college. This was perhaps the happiest period in his life. He was with his loving mother and wife, and he was free to pursue his research. He was now famous all over south India as the amazing genius whom even the great professors in England admired.

But Neville continued to nudge him toward Cambridge, at Hardy's urging. It soon appeared, however, that the ocean was an insurmountable obstacle. Brahmins live by very strict laws that govern everything they do in their lives. Ramanujan was a strict vegetarian, and more than that, he could accept food only if made by a trustworthy fellow Brahmin. In addition, there was a prohibition against Brahmins crossing the ocean. They could travel within India—but not over the sea to England. Many, of course, did go. Britain has had a considerable number of Indian students in its universities, many of them Brahmins, but those who went did so by breaking the rule against travel by sea.

And now, finally, Ramanujan was soaring. His renown increased as he created ever more new mathematical formulas and theorems. He was respected and admired by everyone. What more could he have asked for? Yet one day, Neville came to him and they talked about mathematics, and about England, and to his surprise, he found that Ramanujan was not absolutely averse to accepting Hardy's invitation. He understood that in Cambridge, among prominent mathematicians, he would thrive. He would be able to aspire to loftier goals, to do better mathematics, than was possible in the relative isolation of Madras.

The minor problems were soon resolved: Ramanujan was prepared to break the religious prohibition against travel—promising that he would adhere strictly to his dietary rules even across the seas. But he would not leave India without his mother's blessing. But the mother objected, and the problem was the goddess.

Komalatammal maintained that the goddess Namagiri had told her that she was against her son's leaving India. Ramanujan tried to convince his mother that she had no way of knowing the goddess's wishes, but his mother knew better. After a few more months of wrangling, Komalatammal suggested to her son that perhaps he should try to obtain Namagiri's permission on his own.

And so it was that in December 1913, Ramanujan, accompanied by two friends, set out for the city of Namakkal, eighty miles west of Kumbakonam. Their destination was a rocky promontory on which an important ancient temple dedicated to the goddess Namagiri was perched. After a trek up to the temple, they set up camp on its floor and meditated. Ramanujan and his friends remained at the temple for three days and three nights, and on the third night, Ramanujan had a dream in which Namagiri spoke to him and granted him permission to leave India for England.

According to another version of this story, while Ramanujan did in fact make the pilgrimage to Namakkal, it was his mother who obtained Namagiri's permission for her son to leave India. In this telling, Komalatammal had a dream in which Namagiri told her to allow her son to leave. Dream or no dream, wishful thinking or chicanery, Namagiri seemed to be on board, and Ramanujan was free to pursue his destiny.

On March 17, 1914, Ramanujan set out for England aboard the SS *Nevasa*, of the British India Lines. Janaki had asked her husband to take her with him, but he refused her request, saying that she would be a distraction from his work. Now he was standing on deck, clearly with mixed feelings—excited to go where leading mathematicians could help him advance his mathematical career, but worried about being alone in a strange, cold country. England was indeed cold compared to south India, where it is always warm, even hot, and practically never even cool. And he was also worried about maintaining his strictly vegetarian diet.

Chapter 15

∼

TOGETHER AT LAST

*T*he *Nevasa* made it to the mouth of the Thames almost exactly a month after leaving Madras. Ramanujan disembarked and was taken to London, where he stayed for several days at a center that welcomed Indian students on their arrival in England. Neville arrived home in England around this time, and Ramanujan spent his first few months in Cambridge living with Neville and his wife in their house near Trinity College.

Trinity College Courtyard at Cambridge University

Soon would come the test whether Ramanujan had made the right choice in coming to England. Having worked in isolation his entire life, he welcomed the chance to have a mentor and friend who understood the beauty that he saw in mathematics. And he now had the hope of receiving validation for his years of lonely labor from the top men in English mathematics.

© Springer International Publishing Switzerland 2016
K. Ono, A.D. Aczel, *My Search for Ramanujan*, DOI 10.1007/978-3-319-25568-2_15

One can only imagine the first meeting between Ramanujan and Hardy. Not long after their first conversations, Hardy declared, "He possesses powers as remarkable in their way as those of any living mathematician. His work is of a different category."

Eventually, Ramanujan was assigned rooms at Trinity, very close to Hardy's own rooms. The two men met almost every day for several years, poring over Ramanujan's amazing formulas. Hardy had already received more than one hundred formulas in letters from Ramanujan, and Ramanujan had brought many more with him. Looking at his claims, Hardy could see that many of them were remarkable innovative breakthroughs in mathematics. How could this man have produced them entirely on his own? That was an enigma that Hardy, even with Ramanujan by his side, could not resolve. What does one do with an answer like, "Namagiri, our goddess, presented them to me in my dreams"? Mathematicians would be contemplating the mystery for the next century.

At one point, Hardy described Ramanujan's amazing achievement by saying that in India, he had been working under "an impossible handicap, a poor and solitary Hindu pitting his brain against the accumulated wisdom of Europe." I find this image both deeply moving and highly descriptive of the life of this towering mathematical figure, a unique genius.

In working with Ramanujan, Hardy saw two immediate aims he must try to achieve. First, he wanted to establish complete and rigorous proofs of as many of Ramanujan's claims as possible. The two of them worked together, proving many of the results, and those derivations were subsequently published by the two of them in academic journals. Other results would take many years to prove—by them or by other mathematicians, including my colleagues and myself. A small fraction of the claims turned out to be false. But in those cases, the results still presented interesting methods that could be used to crack other difficult problems and that shed important light on mathematics itself and how it is pursued.

Hardy's second aim was to bring Ramanujan up to speed on modern mathematics. Ramanujan had done all his previous work without the benefit of an advanced mathematical education, like what was available at Cambridge. He thus encouraged Ramanujan to sit in on classes and lectures. That effort was not an unqualified success, because by then, Ramanujan was set in his ways. He had done mathematics according to his own methods and was reluctant to change them. But he did learn how to construct rigorous proofs of theorems.

Chapter 16

~

CULTURE SHOCK

*L*ike my father, who came to America in the late 1950s, Ramanujan had accepted the invitation of a leading mathematician to work in a foreign land. Both men struggled to fit into an alien culture with different languages and customs. In a way, Ramanujan and my father were both fugitives. My father fled the desperate conditions of postwar Japan, and Ramanujan fled a life in which he was intellectually hampered. I believe that my parents responded to the effects of racial prejudice, and in particular anti-Japanese sentiment, by enforcing *kaikin*, isolationism, in our home in Lutherville. Ramanujan, as a foreigner in England, was similarly isolated, separated from virtually everyone apart from Hardy and a few friends. If it were not for their shared absorption in mathematics, I think that both men, Ramanujan and my father, would have had an even more difficult time of it.

Ramanujan would work for hours at a time. Often, he would start late at night, when the world was quiet, and work until dawn. He would then cook breakfast for himself, then sleep much of the day, seeing very little sunshine—both because Britain is notoriously gray and because he largely kept to his rooms. The only other place he frequented was Hardy's rooms, where they would work together. Unlike Hardy and other Fellows of Trinity College who regularly played sports, such as tennis and cricket, Ramanujan's life was completely sedentary.

He was having a difficult time adjusting to life in England. The customs were completely different from what he was used to. The English were formal and cold—unlike the smiling, warm south Indians, ever interested in how you felt and what you were up to. He hated having to give up his long hair for a European cut. The collar of his stiffly starched shirts hurt his neck, and his feet never got used to the pressure of Western shoes. He longed for his carefree life in sunny India, wearing Indian dress, eating the spicy and vibrant food of his childhood, feet unbound.

© Springer International Publishing Switzerland 2016
K. Ono, A.D. Aczel, *My Search for Ramanujan*, DOI 10.1007/978-3-319-25568-2_16

And since as a Brahmin he could not trust the cooking at the Hall at Trinity, where everyone else ate, he was forced to buy his own foodstuffs and cook for himself in his rooms. This had the effect of depriving him of the important social contact that took place in the dining halls of the college, for much of the social interaction among faculty, fellows, students, and others at Trinity took place in the dining halls. Ramanujan was mostly alone.

He worked at night, cooked early in the morning, went shopping for food later in the day, slept, saw Hardy for an hour or two, ate dinner, and worked again. It was an unhealthy life, with few breaks or rewards—other than the mathematics. He had a small gas stove in an alcove in his rooms, and there he would cook his vegetables. He missed the delicious raitas and spicy rice dishes of south India. Here he had to make do with what he could find.

Chapter 17

TRIUMPH OVER RACISM

*L*ike my father, Ramanujan dealt with racial prejudice. Japanese-Americans like my parents suffered the effects of racism in America largely because of World War II. Ramanujan suffered as a dark-skinned Indian living under British imperialism. Both men were able to overcome prejudice through their important mathematical contributions.

Ramanujan at Cambridge (Ramanujan center and Hardy standing at far right)

© Springer International Publishing Switzerland 2016
K. Ono, A.D. Aczel, *My Search for Ramanujan*, DOI 10.1007/978-3-319-25568-2_17

Ramanujan's theorems stunned the British mathematical world. Nothing like his magical legerdemain had ever been seen. There was no question but that he was realizing the potential that Hardy had recognized in his letters.

In March 1916, Ramanujan was awarded a bachelor's degree from Trinity College for his contributions to mathematics made during his time at Cambridge. The doctoral degree was not commonly given for mathematics during that period, and later, when mathematics PhDs began to be awarded, his degree would be upgraded to a doctorate.

Hardy felt strongly that Ramanujan deserved to be elected a Fellow of Trinity College, as he himself had been. So he began to lobby with the other fellows to have Ramanujan thus honored. But he was surprised by the strong resistance that he encountered. He should not have been. To begin with, the relationship between Britain and India was that of colonial power over a conquered people. Part of what makes it possible for one people to colonize another is a belief in the colonizer's superiority. In this case, that sense of superiority had a strong racial element: "Take up the white man's burden" is the notorious refrain of a poem published by Rudyard Kipling in 1899.

As overlords of a subjected people of an "inferior" race, the English rulers of India consciously isolated themselves from the masses they ruled. They lived apart from them; they didn't socialize with them; they worked with them only to the extent necessary. Even the most refined and educated Brahmin could not socialize as an equal with the lowliest British civil servant stationed in India.

Indian students came to Britain in modest numbers, and they were formally welcomed there. But nobody embraced them, just as my father was welcomed to the United States in the 1950s but was not considered a true member of American society.

When Ramanujan's name was proposed for election as a Fellow of Trinity College, one of the English fellows responded that he would "not have a black man as fellow." This defeat depressed Ramanujan greatly: he could be the most gifted mathematician around, but he would be seen in this land as a "black man," unworthy of recognition.

When my father first told me about Ramanujan thirty years ago, he made it a point to stress the injustice that Ramanujan experienced as an Indian in England. That bigotry evoked powerful images of my parents' lives as Japanese-Americans living in the United States in the 1950s and 1960s, and of course, to a lesser extent, it reminded me of the bigotry that I had encountered in my own life.

Another reason why Hardy failed in his attempt to have Ramanujan elected a Fellow of Trinity College had to do with World War I. The "Great War" erupted in 1914, and Cambridge almost emptied of students as they joined the effort to defeat the Germans. Many professors, including Littlewood, also headed to the battlefields of France and Belgium. Hardy remained, as did Ramanujan.

Hardy was opposed to war, even while he understood the necessity to defend Britain and the Continent from German aggression. Although he was not a conscientious objector, he voiced his opposition to how such objectors were being treated. Then at some point during the war, he supported antiwar statements made by the eminent Cambridge logician Bertrand Russell, and that was enough to tar him with the pacifist brush. He was thus politically weakened and could not effectively fight for Ramanujan.

Ramanujan, humiliated and upset by the defeat of his nomination to become a fellow, also suffered physically. It was at this point that the wartime scarcity of fresh fruits and vegetables—the main staples of his vegetarian diet—began to affect his health adversely. He became desperately ill. Naturally heavy, he now lost weight. He talked less, even meeting his only main contact with the world, Hardy, less frequently. One day, in London, he threw himself in front of an oncoming subway car. Miraculously, an operator saw him fall to the tracks just in time to throw the electric switch, cutting power to the trains and thus saving Ramanujan's life.

Despite such adversity, Ramanujan continued to do mathematics—but his health continually declined. He now required frequent hospitalizations, and Hardy began to fear for his life. Hardy then decided that he would try another way to get Ramanujan the recognition he deserved, and so he put him up for election to the Royal Society. If Trinity rejected him, Hardy reasoned, the more hallowed Royal Society might take him. The Royal Society heard the case for Ramanujan, and also considered—something Hardy brought to their attention—that his health was declining and that if he were not elected that year, there might be no future opportunity to recognize his achievements. And so in 1918, Ramanujan became a Fellow of the Royal Society, the second Indian to be awarded this high recognition, and one of the youngest fellows ever. Ramanujan was vindicated.

Ramanujan was also elected to the London Mathematical Society. After that, the Fellows of Trinity College found it hard to say no the second time his name came up, and in October 1918, he was finally elected a Fellow of Trinity College. He had triumphed against prejudice and won the honors he so well deserved.

Chapter 18

~

ENGLISH MALAISE

*E*ngland's unforgiving climate and Ramanujan's poor diet took their toll. Ramanujan's health continued its long decline. He had endured a variety of ailments throughout his life, and after five years in Cambridge, he was constantly ill. The doctors believed he had tuberculosis, although later findings have suggested a parasitic infection affecting his liver. Blood poisoning was also suspected. No one knew exactly what was amiss, but he suffered from fevers, stomach pains, and many other symptoms.

The treatment for tuberculosis was rigorous. Patients were purposely kept in unheated airy rooms, with only blankets to protect them from the elements. The idea was that the lungs could somehow be cleansed through fresh, cold air. In England, he had always suffered from the cold, and now he was to be exposed to cold on purpose. And food—always a concern—now seemed an insurmountable problem, since the institutional food was incompatible with Ramanujan's pledge of maintaining a strict Hindu diet that had to be prepared under Brahmin oversight (namely his own). Ramanujan therefore insisted on doing his own cooking, and after his Indian friends who visited him lobbied the management of the sanatorium on his behalf, he was allowed to prepare his own meals.

Theories abound as to why he became so sick. Tuberculosis can be aggravated, even triggered, by a lack of vitamin D. In England, Ramanujan was always indoors and was therefore exposed to little sunshine, and his diet—mostly fruits and vegetables—did not supply much of that needed vitamin. The privations brought about by the war made things even worse. Moreover, his sedentary nocturnal life was in itself unhealthy. His condition, whatever was causing it, had become acute. He lost weight rapidly. He was in and out of hospital. Finally, even Hardy came to realize that to save his life, perhaps his protégé should return home to India, at least for a while.

© Springer International Publishing Switzerland 2016
K. Ono, A.D. Aczel, *My Search for Ramanujan*, DOI 10.1007/978-3-319-25568-2_18

Chapter 19

~

HOMECOMING

*I*n hospitals, sanatoriums, halfway houses, through declining health, Ramanujan kept up his work. He continued to produce groundbreaking theorems about partitions, identities, infinite series, integrals, and more, though now—in contrast to his work five years earlier—with a proof of every result. He had learned how to do mathematics in the modern style, in which every assertion must be demonstrated using rigorous argumentation.

Ramanujan's health was declining slowly, and that allowed for a host of useless statements to be made about his condition: he is better today, or only slightly worse, or he is on the mend. At some point, Hardy felt that he was doing well enough for a sea voyage, and he suggested both to him and to the academic authorities in Madras, which had cosponsored his sojourn in England, that perhaps it was time for Ramanujan to return home—at least for a visit in hopes of further improving his health.

And the time was propitious. During the war, the German navy had blockaded all the sea routes to Britain in an effort to isolate it both from the Continent and from American aid, which came mostly by sea. With the war now over, sea travel was returning to normal, and it was again a matter of course to book passage to India. But Ramanujan did not express any great enthusiasm for the idea of a sea voyage home. In fact, he was rather cool to the proposal, which Hardy found perplexing. There were several possible reasons for Ramanujan's reluctance.

By now, the young Brahmin had become used to living in England. He had found a way to maintain his diet: buying fresh vegetables wherever he could, cooking them—even in a hospital kitchen—and procuring what he couldn't purchase from Indian friends in England who may have received packages of food from relatives back home and were willing to share them.

Another reason was that he had heard that if as a tuberculosis patient, he was being encouraged to go home, it was a sign that his disease had worsened beyond

© Springer International Publishing Switzerland 2016
K. Ono, A.D. Aczel, *My Search for Ramanujan*, DOI 10.1007/978-3-319-25568-2_19

hope of cure, and he was being sent home to die among his loved ones. There was also a personal reason for his reluctance. He was alone, sick, depressed, and worried that for more than a year, he had not heard a word from Janaki. And for some months, he had heard little from his mother or father. Finally, there was a mathematical reason for his reluctance. Ramanujan was worried that in India, his torrent of mathematical productivity might recede.

For all these reasons, he was in no rush to go home. But it seems that he had little choice in the matter. Hardy was out to save his life and was in contact with Indian colleagues; that arrangements would be made for Ramanujan's return to his homeland—nominally for a visit, now made possible by the end of the war—was a foregone conclusion.

In February 1919, Ramanujan obtained a new passport. The passport photo for which he sat has since become one of the best-known photographs of him. He had all the necessary documents, and passage home had already been booked for him.

Ramanujan's passport photo

On March 13, 1919, Ramanujan paid his last visit to Hardy, left him many of his mathematical papers, and boarded the SS *Nagoya*, bound for Bombay. From there, he would make his way by train to Madras and Kumbakonam.

On April 6, Ramanujan was united with his family in Madras. He was showing symptoms of severe ill health, and he was placed under the care of a physician in a quiet house in the city. Everyone wanted to see him. There were numerous articles about him and his life in the newspapers—he was a celebrity. But he needed peace and quiet if he was to regain his health. It was ordered that the number of people he saw every day be limited and that he be given ample food and rest.

As summer approached, the doctors told Komalatammal that she should take her son to the high country, where it would be cooler and drier than in sweltering coastal Madras. She chose a town in the hills west of Kumbakonam, not far from the hometown that the family had never forsaken while moving around over south India. Money was no longer a problem, for Ramanujan received funding for all his activities, living expenses, and health care from a number of government agencies and other sources. He was one of the most famous people in India, and the biggest problem, other than his health, was to keep people away from him—it seemed that everyone on the subcontinent wanted to see the celebrated genius who had returned home.

Ramanujan spent the summer in a relatively cool environment, working on mathematics. It is hard to describe his health: there were ups and downs, but he was still treated as a tuberculosis patient. His wife would wash his legs and feet, clean the phlegm he would cough up—a sign of tuberculosis, although consistent with other possible ailments—and feed him the foods he had craved for five long years in England: yogurt, rice, pickles, and spicy curries and dal.

Late in the fall, Ramanujan and his entourage moved back to Madras, now that temperatures had dropped to more tolerable levels. By now, he had stopped cooperating with his doctors. He had become an exceptionally willful and feisty patient, frustrating all who tried to take care of him. It is unclear whether he had lost faith in his physicians or simply wanted to be left alone after being shuttled between hospitals and sanatoriums, first in England and now in India. Or perhaps he was now at peace with whatever fate would bring him.

On January 12, 1920, from Madras, Ramanujan wrote his last letter to Hardy. This was the letter that, eighty-five years later, would shape much of my research program. It was his first letter to Hardy in almost a year. Ramanujan began, "I am extremely sorry for not writing you a single letter up to now." And then he dropped yet another of his mathematical bombshells. He had discovered a new type of function, which he called "mock theta functions." Its actual meaning—how Ramanujan would have defined it precisely had he had more space and time—is still debated among experts almost a century later.

Chapter 20

∼

THE TRAGIC END

As Ramanujan approached the end of his life, he was surrounded by domestic strife. His wife and mother were constantly arguing with each other over even the most trivial matters. Anything would send one of them, usually the mother, into a tirade against the other. But Janaki was now twenty years old and no longer a child. She had her husband at her side, who would often support her against the unreasonableness of his mother. Ramanujan's maternal grandmother had also joined the household, and she would frequently add her two cents to the domestic arguments. It was not an environment conducive to healing. Ramanujan would often tell his doctor that he had lost the will to live. There were too many factors wearing him down: his illness, the constant domestic strife, and the self-imposed drive to work on mathematics, especially the new mock theta functions. He was slowly withering away.

Early in the morning of April 26, 1920, he fell into a coma, in which he remained for two hours while Janaki tried to revive him by feeding him milk. He died toward midday.

All of India mourned its national hero, Ramanujan, a man who had brought so much honor and pride to a nation still in the chains of colonial rule. He had triumphed against all odds, he had made a name for himself in England and throughout the world, and he had left a mathematical legacy that would keep me and other mathematicians hard at work over many decades, trying to understand and generalize his results.

After Ramanujan's death, his notebooks and other papers were widely dispersed. George Andrews discovered Ramanujan's so-called lost notebook, forgotten in the Trinity College library, when he visited there in 1976. He spent many years studying it, as did I and other mathematicians. Others of Ramanujan's papers are still being studied, and it will be many years before we understand all

© Springer International Publishing Switzerland 2016
K. Ono, A.D. Aczel, *My Search for Ramanujan*, DOI 10.1007/978-3-319-25568-2_20

that he knew and worked on. Sifting through Ramanujan's work continually turns up nuggets of mathematical gold. The work continues.

Janaki Ammal lived another seventy-four years, until her death in 1994. She adopted two boys, brothers whose parents, who lived very close to Janaki, had died within a two-year period. One of them, W. Narayanan, became a banker, and he remained very close to his adopted mother all of their lives, taking care of her in her old age. Late in life, she would also enjoy financial help from the Indian government—something that had not been forthcoming in the years following her husband's death.

It seems that Ramanujan's mother had taken some of her son's papers and sold them to universities and institutions and that Janaki had received nothing. For many years, she received only a very modest income from her husband's pension. All the rest had gone to his parents. Those who could attempted to exploit Ramanujan's fame for their own benefit. For example, Ramanujan's brother wrote to Hardy asking for money to support his education. It is unfortunate that so little went to Janaki.

Janaki asked very little for herself. In her old age, she simply wanted the broken promise of a statue in Ramanujan's honor fulfilled. Where others had failed, the mathematicians of the world came through, and I am proud that my father was one of the many who helped provide the gift of a bust of Ramanujan. Janaki's touching letter to my father thanking him for his contribution will always remind me of her love for her husband.

Part III
My Life Adrift

Chapter 21

I BELIEVE IN SANTA

Montreal (1984–1985)

My parents dropped me off at Baltimore's Penn Station, and I was now standing alone on the platform waiting for the Amtrak train to New York, where I would change for the "Adirondack" to Montreal. I had my backpack, my Peugeot bicycle, and my large suitcase. My parents were on their way home, and there was no turning back. Like Ramanujan, who ran away to Vizagapatnam after he flunked out of college, I was running away having dropped out of high school. Of course, we had different reasons for running. Ashamed of having lost his scholarship, Ramanujan had simply disappeared. I was fleeing my pressure cooker of a life, but I had left with my parents' knowledge and consent.

It takes over fourteen hours to travel by rail from Baltimore to Montreal, and I passed the time contemplating my past and my future. The metronomic rumble of the wheels as they clicked and clacked over the steel tracks somehow nourished my deflated ego. "Go forth, go forth, go forth," they seemed to say. "Have hope, have hope, have hope." I told myself that I was setting out to live a life that was true to myself, that I was the master of my destiny, and soon my somber mood was replaced by optimism and excitement. I thought about my friends, and I wondered whether I would ever see any of them again. By the time the train pulled into New York, I was thinking about what it would be like to live in a French-speaking city. Would the radio stations play the songs I knew? What would it be like to ride the metro? Whom might I meet? Would I make friends? I had no idea what the future might bring, and I was oddly excited by that uncertainty.

Now, Montreal was not a random choice of destination, and there was one certainty that comforted me: my brother Santa would be waiting for me on the

© Springer International Publishing Switzerland 2016
K. Ono, A.D. Aczel, *My Search for Ramanujan*, DOI 10.1007/978-3-319-25568-2_21

platform in Montreal. Santa had somehow managed to escape the formula that my parents had laid out for him. Perhaps because they had such relatively low expectations of him, he had been granted greater scope in finding his own way. After earning his bachelor's degree in biology from the University of Chicago, Santa immediately began to work toward a doctorate in biochemistry and immunology in Montreal, at McGill University. My parents had mapped out for him a career in industry or business, and so we were all surprised when he decided to pursue a career in academia. As it turned out, perhaps unsurprisingly to those readers with more liberal ideas of childrearing, Santa at twenty had a better idea of his strengths and interests than his parents had had for him when he was six. Santa would go on to hold faculty positions at Harvard, Johns Hopkins, University College London, and Emory University, and he is now the president of the University of Cincinnati.

Five years apart in age, we had never really been close, and so I was touched by his offer to look after me, and I was looking forward to living with him. But it turned out that Santa had other plans. To my surprise, he had arranged a room for me at the Alpha Delta Phi fraternity, on Rue Stanley, in downtown Montreal. I was totally unprepared for so much independence. As the sixteen-year-old son of tiger parents, I had experienced only a very narrow swath of life outside of home and school. I had little or no practical skills or street smarts. I was also unprepared for looking after myself. I had never even had my hair cut by anyone other than my mother.

But I moved in, stowed my bike, put away my things, made my bed, and managed to get through the first night without becoming completely unnerved. But the next morning, when I opened the door of the common bathroom, I was confronted with the sight of an unbelievably hairy young man's back. He was enormous. Wearing nothing but a Budweiser bath towel wrapped around his capacious waist, he was holding a Walkman in his left hand while brushing his teeth with his right. His head was bobbing to the beat of whatever was blasting from his headphones. Sensing that he was no longer alone—perhaps he saw me in the mirror—he turned in a flash, revealing a pair of gold aviator sunglasses. At the sight of me, he yelled, I suppose in order to be heard over whatever music was pounding into his head, "Kid! Cool, man! Give me five!" He turned out to be a football player on McGill University's club team. Nice guy, I suppose, but he scared the hell out of me.

I lived in the frat house for a few days, cowering in my room, wondering what sort of "Animal House" I had gotten myself into.

Santa responded with good grace to my terror by offering me the sofa in his one-bedroom apartment. Santa, of course, knew all about my insecurities, and all about the voices in my head that continually spoke to me of my inadequacy. We had, after all, grown up in the same tiger cage, and he understood me completely. He told me that he heard the same voices. For the first time in my life, I felt that I had someone in whom I could confide, someone who understood my inner turmoil and could offer the loving and nurturing care that I so sorely needed.

Although we had not been close before, a strong fraternal bond formed between Santa and me during my stay with him in Montreal. Through him, I encountered much of the outside world that had been forbidden to us in Lutherville. He taught me how to lead an independent life, and he supported my adventurous spirit. He took me for my first professional haircut. He taught me how to use the metro and how to navigate a large and bustling city safely on my own. He encouraged me to join Team West Island, a local cycling club. I participated in a few races with them, most notably the Gran Prix Lachine, in which we raced around the lovely Parc Lasalle, with a hyperenthusiastic announcer who referred to me as *le jeune Américain*. We played softball for his department's intramural team. We went to the cinema, where we saw the Michael J. Fox film *Back to the Future*. How I enjoyed not having to write an essay about it! We ate late-night souvlaki at a restaurant called Kojax. I went with Santa and some friends from his lab to my first rock concert, where we saw Corey Hart perform his huge hit *Sunglasses at Night* at the Montreal Forum.

I have many fond memories of the time I spent in Montreal, and I am certain that I would not be where I am today had I not run away from my former life to live with my brother. It was thanks to Ramanujan that my parents had let me go, but it was Santa who set me on my feet and helped to point me in the right direction.

Santa had arranged a part-time job for me working in his lab, which gave me an important degree of stability. But I was primarily interested in exercising my newfound freedom as a teenager living away from home for the first time. Sneaking into Thomson House, a limestone mansion on the McGill campus where graduate students gathered for drinks after work, was my favorite pastime. We hung out in its stately rooms, and it was there that I developed a taste for Canadian beer: a bottle of Molson Golden or Labatt Blue can still evoke happy memories of my time in Montreal.

During my time in Montreal I was transformed. I developed a relationship with my brother Santa that was closer than any I had ever experienced. He helped me to find my self-confidence, and he taught he how to put fun into my life. He prepared me for life outside our kaikin home in Lutherville, for life in the real world.

Eight months after my father received the letter from Ramanujan's widow, I was ready to apply to college. I sent applications to Johns Hopkins, MIT,

Princeton, and the University of Chicago. Getting into college was much easier in the 1980s than it is today, and I was armed with strong SAT scores (700 verbal and 800 math) and a recommendation from Johns Hopkins psychologist Julian Stanley, the man who had devoted his career to studying gifted and talented youth. Despite my status as a high-school dropout, I was accepted by Johns Hopkins and the University of Chicago. Given a choice, there was no way I was going to go to Hopkins. It was much too close to home. So I accepted the offer from the University of Chicago, Santa's alma mater. The summer passed pleasantly and uneventfully, and in September, my parents drove up to Montreal, and we headed west for Chicago.

Santa's wedding day in 1989 (*left to right*: Santa, Wendy Yip-Ono, Takasan, my mother, Momoro, Ken)

Chapter 22

COLLEGE BOY

Chicago (1985–1988)

I arrived at the University of Chicago in September 1985. Its campus, constructed largely in the collegiate Gothic style, occupies a two-hundred-acre site on Chicago's south side. The northern and southern parts of the campus are separated by the Midway Plaisance, a broad mile-long park that was constructed in 1893 for the Columbian Exposition. Whether by chance or design, I was assigned to Santa's former residence, Burton–Judson Courts, just south of the Midway. I had a room in a first-floor corner suite in Dodd–Mead House, one of the six houses that constitute Burton–Judson. The neo-Gothic architecture, the leaded windows, the view across the Midway to Harper Library, all enhanced my already overbrimming confidence in my success as a college freshman. Several of the older students remembered Santa, and I was delighted to learn that my hero astrophysicist Carl Sagan had lived in the same dormitory in the 1950s, a few doors from my own suite.

The University of Chicago has a long history of admitting students like me. Each year, in fact, Julian Stanley would recommend a handful of high-school dropouts to the university. This year, he had also recommended two brothers from a Chicago suburb, aged ten and eleven. At seventeen, I was clearly not going to be viewed as an "underage" student. The younger brother was a mere pipsqueak of a kid who wore a Pac-Man watch and was so small that his feet dangled well above the floor as he sat in class, swinging his feet wildly like any other restless fourth-grader.

Founded in 1890, the university has risen into the ranks of the world's top universities. It claims eighty-nine Nobel laureates among its faculty and alumni, more than any other institution of higher learning in the world. Many of the world's leading thinkers have been on the UChicago faculty, including T.S. Eliot,

© Springer International Publishing Switzerland 2016
K. Ono, A.D. Aczel, *My Search for Ramanujan*, DOI 10.1007/978-3-319-25568-2_22

Enrico Fermi, Hardy's friend Bertrand Russell, to name just a few. I felt myself part of a glorious tradition, and I was ready to sally forth and add my name to the college's roll of honor. At the very least, I was going to study hard, do well, and not screw up.

My plan, if you could glorify my act of desperation with such an epithet, had worked. I had escaped from high school, and I was somehow given the chance to start life anew, at a first-rate university, with great expectations. It didn't matter to me that I was a dropout, or that I would have to repeat a grade for failing all my high-school courses if for some reason I had to return to Lutherville.

My parents were worried about me, as they always were, and they were justified in their concern. After all, I had disappointed them in one way after another. But I promised them that I would take college seriously. I even promised to quit bicycle racing. In fact, I left all of my bikes behind. I assured them I would study hard and earn good grades. I felt that I was ready, and I began college with high hopes. I wanted to prove to my parents that by living my life my way, not by their formula, I could succeed.

Alas, I had built my house upon the sand. I had no idea that it would be a mighty struggle to transition from the carefree life I had enjoyed in Montreal to the hardcore labors of a University of Chicago undergraduate. Actually, I didn't struggle. I didn't really transition at all. Instead, I reveled in my freedom, and I made the most of new social opportunities and the great city of Chicago. I was no longer one of the few Asian-American kids at school; there were many of us in my freshman class. I was making new friends, and I was focused almost entirely on enjoying every moment of my new life.

Many first-year college students have difficulty adjusting to their sudden freedom from parental supervision, but I suspect that the children of tiger parents are at an even greater risk of crashing and burning. You would think my year in Montreal would have eased the transition, but I guess I had a lot more pent up in me that was waiting to burst out. I lived recklessly, and today, as a professor at Emory University, I witness the same sort of behavior among each new crop of first-year students. College life offers all sorts of forbidden fruit, and many kids overindulge, as though there were a biological imperative to make up for lost time.

And boy, did I make up for lost time! I was a total goofball. I went to lots of parties. I lived for Harold's fried chicken, a super-tasty, cholesterol-laden delight smothered in hot sauce and barbeque sauce, which was prepared in enormous deep fryers behind a thick wall of bulletproof glass. I was a regular at Medusa's, an underground dance club near Wrigley Field that was popular among under-age college students.

My predetermined path was mathematics, so I of course rejected it, deciding to take premed courses instead. Two months into the quarter, I took my first exam in honors chemistry. I received a C+. I suppose the A and B students were studying more and partying less. Staring at the exam as I walked back to Burton–Judson in the dark, I was in a state of shock. I had never received such a poor grade on an exam before. In fact, I don't think I had ever received a grade as low as an A–. I had the urge to run, at once, and as far away as possible. But I couldn't run anywhere. The campus was enclosed by dangerous neighborhoods on three sides and Lake Michigan on the fourth. That mediocre grade deflated my self-confidence. I was no longer enjoying college life.

Then I got an F on my first paper in my common core course in sociology. The professor, a distinguished economist, began his lengthy comments, written in damning red ink, with the words, "It is obvious to me that you do not understand much of Marx at all." High-school classes had been a breeze. UChicago classes were putting up a strong headwind. My professors were respected leaders in their fields, and they all had high expectations. The voices in my head were ready to chime in with their condemnation: "Ken-chan, you no good. You not study hard enough." Out of those voices I was finally able to compose a simple syllogism: if you don't want to be a failure, Ken-chan, you had better apply yourself.

Like Ramanujan, I struggled with college, although both of us had the intellectual ability to earn good grades. Each of us, however, suffered from an addiction. Ramanujan was addicted to mathematics and was achieving important results, while I was addicted to goofing off and having a good time and was not achieving much of anything.

Although I knew that if I applied myself to my courses, I could do well, I did my best, immature lad that I was, to avoid having to come to terms with anything whatsoever. Instead of working harder, I simply abandoned premed, switching my major to mathematics. I was not happy about returning to my predestined path, but I figured I could get by much more easily in math than in the other majors offered at UChicago, despite the fact that I was behind the other math majors in coursework, including the two prodigious brothers. Most UChicago math majors entered college having taken at least one year of calculus. I, on the other hand, having dropped out of high school, had never taken any calculus at all. Of course, I would have taken calculus in high school had I not dropped out, so I didn't feel quite like a dummy. But I knew that I had some catching up to do.

My speculation turned out to be correct. I was able to earn respectable grades in my math courses without exerting much effort, and that was important, because at the time, effort was not my strong suit. I had a knack for learning proofs, and that was enough to get by.

I took courses from distinguished mathematicians such as Israel Herstein and Raghavan Narasimhan. I did reasonably well in their classes, but I was not a star student. I wasn't even an excellent student. It was as though I was seeking medi-ocrity. I didn't want to compete and be recognized as someone exceptional. Instead, I sat in the back row of my math classes seeking anonymity. I didn't want to return to my high-school role as the nerdy Asian-American kid.

In the spring quarter, I took Math 175, Introduction to Number Theory, from Professor Herstein. The two prodigious brothers were also in that popular class, and they sat next to each other in the front of the classroom while I skulked in the back row. I had enrolled in the course because I had heard about the legend-ary Israel Herstein years earlier from my father. Unfortunately, it turned out that Herstein was seriously ill with cancer. Although he hid the fact of his illness, he was not himself, and he probably should not have been teaching the class.

He would shuffle back and forth behind the lectern, taking drags of a cigarette between sentences. He wrote almost nothing on the board. Instead, he would lecture by bombarding us with questions in an angry tone. He was so intimidat-ing that almost no one ever answered. There was certainly no way that I was going to volunteer, even when I knew the answer, for fear of resurrecting my old image as an Asian-American math nerd. In class after class, he bombarded us with his hostile questioning, and we sat, silent. Out of frustration, he would throw chalk at us, even saying one day that he could "teach this stuff to mon-keys." Although I did fairly well in his class, my negative experience blinded me to any possible beauty in number theory. I would continue to do mathematics, but there was no way that I was going to like it. My plan was to earn decent grades while putting forth minimal effort. My goal was to get by in my courses while leaving enough time for my first priority, my social life.

Although, as promised, I wasn't racing my freshman year, I began doing some cycling. Having left my Peugeot in Lutherville, I had to obtain a bike, and I lucked out. A friend in Burton–Judson lent me a decent Peugeot, which he let me keep in my room. I went for long rides on the Chicago lakefront. On one of those rides, I met Tom Kauffman, a fifty-five-year-old security guard at the Illinois Institute of Technology who was one of the area's top racers in his age group. He often invited me to join him on his rides, teaching me the circuitous routes that offered safe passage through the tough neighborhoods of Chicago's south side. Tom, who was a high-school dropout like me, had been a cook in the U.S. Navy. He became one of my confidants, a father figure who nurtured me outside of my college life. I rode a good deal, both with and without Tom, and I was soon back in good form.

And it came to pass that I got through my first year at UChicago, if not with flying colors, at least not with my tail between my legs. I spent the summer back in Montreal, working in my brother's laboratory cleaning test tubes.

On returning to Chicago in the fall, I soon realized how much I missed bike racing. So after a year of keeping my no-racing pledge, I went off the wagon, without telling my parents. It wasn't long before I was invited to race for the South Chicago Wheelmen, a local club sponsored by Schwinn Bicycles and the Chicago Dough Company, a local Italian restaurant. I was passionate about cycling, but I had no idea that I would one day be just as passionate about mathematics.

Tom and I had become quite close, spending hours traveling, odd couple that we were, to races all over the country. He would race in his age group, while I was learning the ins and outs of competing in the senior eighteen to thirty-four contingent. Tom won almost every race he entered, while I struggled to perform in races that were much longer and faster than those to which I had been accustomed in the fifteen-to-seventeen age group. Thanks to Tom's patience and mentorship, I advanced through the ranks and became an accomplished cyclist. I knew Tom for only a few years before his life was tragically cut short when he was killed at work, the victim of a stray bullet. I still grieve over my loss, and I am thankful for the time we spent together. Tom taught me that you didn't need an advanced degree to live a happy and fulfilled life.

My first pedal stroke of the 1987 Mid-America Time Trial Championships

Cycling helped me in unexpected ways. The exercise, of course, was good for my health. If I was going to have a sound mind, it needed to be housed in a sound body. But the most important benefit was the strength of character I built from the regimen of long daily training rides in unforgiving conditions, including dodging cars in Chicago traffic and negotiating unfriendly neighborhoods, a daily real-life version of the classic Disney film *Mr. Toad's Wild Ride*. My wild wintry rides were often miserable, thanks to ice, snow, and the piercing bone-chilling winds off Lake Michigan. I have painful memories of frostbitten cheeks and numb fingers and toes on those early-morning rides to Sauk Village, Illinois, twenty-five miles to the south. The return trips were mighty struggles against freezing headwinds in a desperate race to get back in time for my morning classes, a race that I often lost.

That fall, I joined the Psi Upsilon fraternity, a fun-loving crazy group of guys who lived in an old house across from Bartlett Gymnasium on University Avenue. The house had been around since 1917 and had acquired a permanent odor of stale beer and Pine-Sol, and its creaky stairs did little to conceal its age.

Psi Upsilon fraternity house (photo by Doug Jackman and Chuck Werner)

I was proud to be a brother, to be part of a lineage that included Supreme Court Justice John Paul Stevens and football star Jay Berwanger, the first to be awarded the Heisman trophy. To think that only a year before, at the beginning of my time in Montreal, I had been terrified of living in an "Animal House," and now I was a card-carrying member of an Animal House and loving it.

I never told my parents that I had joined Psi U. It was one more thing that I hid from them, just as I had hidden my seventh-grade math tests. I had no desire to subject myself ever again to their withering criticism. As a brother at Psi U, I became a respectable foosball player. I also became a DJ, following in the footsteps of my friends Joe "Spike" Melendres and Harold "Hoda" Tsai. We spun twelve-inch singles at campus dance parties. New Order's *Bizarre Love Triangle* and Dead or Alive's *Brand New Lover* were my favorites.

Preparing to deejay a Psi U Halloween party (*left to right*: Spike in costume, Ken Ono) (photo by Joe Melendres)

Chapter 23

∽

ERIKA

I had moved out of Burton–Judson at the end of my freshman year and was now living in Shoreland Hall, a building dating from the 1920s, when it opened as the luxury Shoreland Hotel, a place where Al Capone held "business" meetings and Jimmy Hoffa kept a room. Another famous resident was the economist Milton Friedman. But nothing lasts forever (except eternity), and by the 1970s, the Shoreland had come down in the world considerably. Needing another dormitory, the university bought the hotel and converted it into a residence hall. That incarnation lasted until 2009, when it was sold to a private developer and renovated into luxury apartments. I had a room on the sixth floor with a breathtaking view of Lake Michigan.

One of the resident heads in Shoreland had a cherubic, bubbly three-year-old daughter named Kelsey. Whenever Kelsey and her mom ate dinner with us in the cafeteria, Kelsey was the center of attention. We all loved her, and there wasn't enough of her to go around. But Kelsey had a favorite, a freshman girl from Montana with green eyes and wavy blond hair. Whenever Erika was around, you were likely as not to find Kelsey in her lap.

Many of my earliest memories of Erika Anderson, who in 1990 would become Erika Ono, are related, if a bit tangentially, to cycling. I would breakfast at Pierce Dining Hall on Saturday mornings before almost anyone else was awake. I wanted to get an early start to my long training rides into northern Indiana. Erika, who was a freshman in 1986, had drawn the morning shift at Pierce as part of her work–study job. Despite the early hour, she always had a smile on her face, a smile that I looked forward to seeing week after week. She made me delicious omelets, and she smuggled bunches of bananas to sustain me on my long rides. She has been sustaining and nurturing me ever since.

© Springer International Publishing Switzerland 2016
K. Ono, A.D. Aczel, *My Search for Ramanujan*, DOI 10.1007/978-3-319-25568-2_23

Erika, the love of my life, is a gift from Missoula, Montana. Missoula is a geological oddity; it is located at the convergence of five mountain ranges and the confluence of three rivers, and as its topography suggests, it is a place rich in natural beauty, some of which, as I think I have suggested, had recently migrated to Chicago. Lewis and Clark were the first European explorers to visit the area, and it is the setting of the classic autobiographical novel *A River Runs Through It*, a tale of two brothers coming of age during the Great Depression and the Prohibition era.

Unlike my family, in which the focus had always been on intellectual achievement, Erika's family emphasized spiritual and physical communion with nature. Missoula was the perfect place for their active lives. Erika's father, Robin, who taught high-school biology for thirty-five years, is an all-around athlete and outdoorsman. He is a fly fisherman, Nordic skier, runner, and windsurfer. Erika's mother, Jan, who was an elementary-school librarian, was a national-class downhill skier in high school. She played soccer and ice hockey into her sixties. Erika emerged from their parenting as the adventurous, fun, maternal, thoughtful, and warm person that has blessed me throughout my adult life.

With Erika at the 1987 Psi Upsilon Valentine's Day formal

And it is not only I who have been blessed by Erika's presence. Although she studied French literature at the University of Chicago, it was there that she discovered her true passion: babies and their mothers. She would later earn a second undergraduate degree in nursing, and then a master's degree in nurse midwifery from the University of Pennsylvania. She loves helping women discover their strengths as they become mothers. She has welcomed hundreds of babies into the world, two of them ours.

Chapter 24

$\sim$

THE PIRATE PROFESSOR

Chicago (1988–1989)

*U*Chicago's Department of Mathematics is located in Eckhart Hall, a distinguished neo-Gothic building on the main quadrangle across the street from Alpha Delta Phi, a rival fraternity. Many famous mathematicians have taught in this building, including André Weil, the man who discovered my father in 1955 in Tokyo. The department's history and the imposing nature of the building intimidated me from day one.

Despite my desire for anonymity, it was soon apparent that a math major named Ono had nowhere to hide. It was well known among the math faculty that I was the son of the famous Takashi Ono at Johns Hopkins. Professor Walter Baily, a senior number theorist and family friend, showed great kindness to me by inviting me to his home for dinner my freshman year. He welcomed me to UChicago, and as a family friend, he wanted me to know that he was always available in case I ever needed anything. Here was a chance for a friend and mentor in my major field. But stubborn fool that I was, I never spoke with him again. I didn't want to be known as Ono's son, and I was too insecure to feel that I deserved the goodwill of such an eminent mathematician. My failure to accept his friendship and mentorship haunts me to this day. Even long after I finally grew up and figured out who I was, I never took the opportunity to reconnect with him. And now it is too late. Professor Baily passed away in 2013.

My inability to remain anonymous was disconcerting. Not only was I expected to be a star simply on account of my name, I felt that if I fell short, my failure would not be mine alone; I would be disappointing and embarrassing my father.

© Springer International Publishing Switzerland 2016
K. Ono, A.D. Aczel, *My Search for Ramanujan*, DOI 10.1007/978-3-319-25568-2_24

I had put half a continent between myself and my parents, yet despite the distance that separated us, I still heard voices that hammered at my self-esteem:

Ken-chan, you no can hide. Your professors know family, so it your duty to live up to name of Ono. You must be one of best, and right now you losing out to ten-year-old kid with Pac-Man watch.

I didn't care enough to apply myself to my coursework. I was having too much fun spinning records at Psi U and grinding the gears of my bicycle in races all over the midwest. I stubbornly rebelled against my parental voices. I wasn't going to let them tell me what to do. It's not as though I was still the goofball screw-up of my freshman year. I was taking difficult math classes, and I was earning acceptable grades. To my frat brothers and cycling friends, I was some kind of math god merely based on the names of the courses I took, which included Differentiable Manifolds, Algebraic Topology, and Number Theory. I wasn't a brilliant math major, but I was doing well among a cohort of strong students. It didn't seem fair that the voices in my head continued to speak relentlessly to me about my inadequacy.

But I had finally to admit that in some sense, perverted though it may perhaps have been, there was some truth to what those voices were saying. The best math majors at UChicago were extremely talented, and many first-rate majors maximized a less-than-extreme talent through hard work. Without talent and a strong work ethic, you were not going to be a star student at UChicago. In this regard, if I entertained thoughts of being "among the best" at UChicago, then it was true that I was inadequate. I was talented, but I was no genius. You can be a talented composer and work hard at your craft, and then along comes a ten-year-old Mozart and leaves you in the dust. You can be a talented number theorist and labor over theta-function identities, and then along comes a Ramanujan, and you realize that there are minds that pull this stuff out of a dimension to which you will never have access. To be sure, I could have done much better than I was doing, but I wasn't willing to put up with the late nights of problem sets. I wanted to enjoy my social life in Psi U, and I wanted to race my bike on weekends. And furthermore, if I gave my studies my all and still came up short, my voices would triumph over me. For now, I could tell them, "Of course I am inadequate, but that's because I'm not really trying."

I didn't know how to work hard enough to make the most of my talent, and I didn't have the will to find out how. It would take many years before I would be able to come back that short distance correctly. So I told the voices to go to hell. I rejected any desire to earn the praise and approval of my parents. I had decided that their goals for me would not be mine. But without their goals, I had no long-term goals at all. I was just marking time.

Then a shocking incident occurred at the end of my junior year. My instructor in complex analysis, who was a junior visiting professor, summoned me to his office on the second floor of Eckhart Hall. The course requirements included three midterms and a final exam. I had earned a high A on the first midterm and a low A on the second one. On the third exam, I think I earned a B, a sign that I hadn't compensated for the increasing difficulty of the subject by increased effort. But I had plenty of points from the earlier exams. In fact, I had an A-minus average, so why was I being summoned? What did this instructor want?

I had never been to his office, and so I was surprised by the sight of his stark, barren room when I found it. Apart from a few books and some papers, the office was nearly empty, and it was covered in years of dust. There were almost no personal belongings. Here was someone who apparently lived entirely in his mind. All he needed was a desk and chair.

He told me that he knew my father and that he felt compelled to tell me that from what he had seen of my work, I was unlikely to be successful as a professional mathematician. If that was what I was aiming toward, then I was wasting my time chasing a hopeless dream. He encouraged me to pursue some other career, one that didn't involve proving abstract mathematical theorems. He had concluded that I didn't have the talent to make it as a theoretical mathematician. But that was not to say that I had no gift for working with numbers. Perhaps I should consider a career in finance or banking.

I was stunned. I couldn't believe what I was hearing. Here was a real live person confirming the voices in my head. Everything that he was saying added up to one simple sentence: "Ken-chan, you no good." I was so angry that it took all my self-possession to keep from shouting at him. He must have assumed that I was working hard in his class, giving it my all, when in fact, I was putting in hardly any effort. How dare he tell me that I wasn't up to snuff!

Although I laughed the incident off with friends a few hours later over beers at the frat house, I was still burning with rage. I knew that I had no intention of becoming a mathematician, but who was this interloper to tell me that I didn't have it in me to become one if I so chose? On the other hand, it was true that I was performing at a level below my ability. If that professor had known how little effort I was putting into his class, he would have been right to question my maturity and work ethic. But where did he get off questioning my ability, when I hadn't even begun to show what I could do if I tried? The way I saw it, that professor had come to regard me as the struggling son of a famous mathematician who didn't have a prayer of living up to his father's expectations.

I wasn't going to take that without hitting back! Although my first reaction was one of anger, what followed was a mixture of pride and defiance. No random professor was going to tell me what I was or wasn't capable of doing. It was one thing to rebel against my parents and the voices in my head, but this stranger had put my back up. I am sure that he thought he was offering me sound advice. But I was going to show him! I vowed that I would make some serious changes in my senior year and turn things around. I still held out little hope that I would ever drown out the negative voices in my head, and I didn't believe that I could ever earn the approval and respect of my father, but nevertheless, I would set a lesser goal and show those professors that I could work hard and be a strong student, one worthy of their respect.

I spent the following summer racing my Italian Basso bike and interning for a Chicago actuarial firm. While watching TV one evening in my room at Psi U, I encountered the BBC documentary *Ramanujan: Letters from an Indian Clerk*. It was four years since I had thought about Ramanujan, and hearing his uplifting story again gave me hope. And this time, the story was on screen, in color, and I was mesmerized by the footage of Cambridge, England, and villages in south India. I was fascinated by interviews with the distinguished scientists George Andrews, of Penn State, who would later become my colleague; Béla Bollobás, of Cambridge University; and the Nobel-laureate physicist Subrahmanyan Chandrasekhar, of UChicago.

Thanks to Ramanujan, my parents had allowed me to run away from my former life. Four years later, as a rising college senior, uplifted by the documentary on Ramanujan and goaded by the realization that I had only three quarters of college left, I vowed to give my coursework my full attention. I was no longer ashamed of being an Asian-American math nerd and the son of tiger parents, one of whom was a famous mathematician. Thinking again about Ramanujan helped me appreciate my knack for numbers. I decided to embrace my identity as an Asian-American tiger cub with some talent for math, and I prepared to apply myself in my courses my senior year. I would prove to the UChicago math faculty that I had real ability, that I should be taken seriously. I would worry later about what I would do with the rest of my life.

I moved out of Psi U, and I didn't go to many parties my senior year. I devoted myself to mathematics, Erika, and bike racing. I was invited to race for the Pepsi-Miyata cycling team, one of the best teams in the midwest. We raced on state-of-the-art Miyata bikes, and our skinsuits made us look more like giant cans of Pepsi Cola than super-skinny cyclists. My main traveling partner was Ron

Lynch, who would later be a groomsman at my wedding. We crisscrossed the midwest in his Subaru, traveling to races with U2's *Joshua Tree* blaring on the stereo. The 1989 season, my last as a competitive cyclist, would end for me on Father's Day in Philadelphia at a place called Manayunk, where I got the chance to race three-time Tour de France champion Greg Lemond in person in the Core States Cycling Championship. It was an awesome experience.

Racing up the Manayunk Wall in 1989 (photo by Dave Mathis)

I enrolled in challenging math courses my senior year, and I made it a point to stand out. I sat in the front row. I did all the homework. I spoke up often in class. I still wasn't passionate about the mathematics, but I enjoyed the newfound attention from my professors. I took one particularly challenging class from Professor Paul Sally, who was known as the "Pirate Professor" because of his towering figure, colorful personality, and eye patch.

Sally suffered from adult-onset diabetes, and his body slowly betrayed him over the twenty-five years that I knew him. He died unexpectedly on December 20, 2013. By the time I was a student in his analysis class in 1988, he had lost the use of his left eye, whence the eye patch. Later, he would lose the use of his legs, first one, then the other, being finally left wheelchair-bound as a double amputee. In the last years of his life, he was almost blind. Despite all of his health problems, he was a dominant figure with a booming voice at UChicago. He had a take-no-prisoners approach to undergraduate mentoring. He sensed that there was something in me that needed nurturing, and he was there to help me find my way.

"Pirate Professor" Paul Sally (photo by Sharat Ganapati)

Shortly after Halloween, he called me into his office, which was actually a spacious suite of rooms on the second floor of Eckhart Hall. Nature apparently abhors a vacuum more in some places than in others, and this was one of them. Almost every square inch of Professor Sally's suite was covered in mountain ranges of books and escarpments of papers stacked high on the floor, on desks, on tables, in a geological formation that must have taken him decades to create. There was just room for a fridge in the corner and access to a super-dusty chalkboard covered with formulas and Sally's weekly schedule.

It was just the spot for a heart-to-heart conversation. Sally told me that I was one of the best students in his class, which also included some UChicago graduate students, and so he wanted to chat about my future. He did this with many of the math majors at UChicago. Sally was a world-class mentor, serving for decades as the department's director of undergraduate studies.

I told him that I didn't have much fondness for mathematics, and I began to relax when he didn't reproach me. He said that whether I liked math or not, I appeared to have a talent for it. He mentioned my uneven performance in UChicago's math classes, of which he had firsthand knowledge, and he told me that he respected the fact that I was an avid bike racer. He must have known about that from a few articles about my cycling in the *Chicago Maroon*, the campus paper. Then, to my surprise, he told me that we had a lot in common. His first passion, he said, hadn't been mathematics; it was basketball. He had been a star player for Boston College High School in the 1950s. He, too, had pursued a

COLOR PLATES

Ramanujan in 1912 at the University of Madras.
Photo by R. Shankar

With Erika on Maui (May 27, 2015)

My second-grade portrait

My parents in 1999 (photo by Olan Mills)

Takasan, ca. 1935

Part of the group photo at the 1955 Tokyo–Nikko conference (*Top row*: second from left, Jean-Pierre Serre; third from left, Yutaka Taniyama. *Third row*: second from left, Takashi Ono. *Second row*: on left, Goro Shimura. *Bottom row*: third from right, André Weil)

Anti-Japanese sentiment in the 1940s

Fuld Hall at the Institute for Advanced Study (photo courtesy of the Institute for Advanced Study)

Goofing off before a race

Ono Family mid 1980s (*left to right*: Takasan, Ken, Santa, Momoro, my mother with Igor)

S. Janaki Ammal
W/o (Late) Srinivasa Ramanujam
(Mathematical Genius)

S. Ramanujan
1887-1920

45, Muthiah Mudali IInd Street,
Krishnampet, Madras-600 014.

INDIA

Date 17.3.1984

Dear Sir,

I understand from Mr. Richard Askey, Wisconsin, U.S.A., that you have contributed for the sculpture in memory of my late husband Mr. Srinivasa Ramanujan. I am happy over this event.

I thank you very much for your good gesture and wish you success in all your endeavours.

Yours faithfully,

S. Janaki Ammal

Janaki Ammal's letter

Janaki Ammal in 1987

Granlund's bust of Ramanujan

Map of South India (drawn by Aspen Ono)

Ramanujan's mother, Komalatammal

Janaki Ammal

G. H. Hardy

13

Salvador Dalí's *The Sacrament of the Last Supper* (courtesy of the National Gallery of Art, Washington, D.C.)

Trinity College Courtyard at Cambridge University

Ramanujan at Cambridge (Ramanujan center and Hardy standing at far right)

Ramanujan's passport photo

Santa's wedding day in 1989 (*left to right*: Santa, Wendy Yip-Ono, Takasan, my mother, Momoro, Ken)

My first pedal stroke of the 1987 Mid-America Time Trial Championships

Psi Upsilon fraternity house (photo by Doug Jackman and Chuck Werner)

Preparing to deejay a Psi U Halloween party (*left to right*: Spike in costume, Ken Ono) (photo by Joe Melendres)

With Erika at the 1987 Psi Upsilon Valentine's Day formal

Racing up the Manayunk Wall in 1989 (photo by Dave Mathis)

"Pirate Professor" Paul Sally (photo by Sharat Ganapati)

Newlyweds (photo by Jan Anderson)

With Basil Gordon in 1992 (photo by Keith Kendig)

George Andrews in 2008 (photo courtesy of the American Mathematical Society)

Bruce Berndt with Ramanujan's slate in the 1980s (photo courtesy of Bruce Berndt)

In Lutherville after the Rademacher conference (*left to right*: Doug Bowman, Ken Ono, Takasan)

With Andrew Granville in 2013

André Weil lecturing at the 1955 Tokyo–Nikko conference

In front of the White House before my Presidential Early Career Award ceremony

Sound and Manjul in front of Ramanujan's boyhood home (photo by Krishnaswami Alladi)

The Sarangapani Temple (photo by Krishnaswami Alladi)

COLOR PLATES

Perusing Ramanujan's notebooks with Raghuram (photo courtesy of Emory University)

With Jeremy Irons on the set (photo by Sam Pressman)

Preproduction party (*left to right*: Edward Pressman, Jeremy Irons, Dev Patel, Ken Ono, Matt Brown, Sorel Carradine) (photo by Sam Pressman)

George Andrews, Michael Griffin, Ken Ono, Ole Warnaar, Jim Lepowsky

With Ramanujan's last letter to Hardy in 2013 at Trinity College

John Duncan, Ken Ono, Michael Griffin

Ken Ribet, Peter Sarnak, Ken Ono in 2014 (photo by Ling Long)

Ken Ono and Dev Patel rehearsing the "Circle Method Scene" (photo by Sam Pressman)

number of interests during his undergraduate years, and had been unsure about a future in mathematics. However, at some point, something sparked his desire to pursue mathematics, and he somehow found his way, his journey eventually taking him to a professorship at UChicago. He mentioned that for a brief stretch, he had worked as a Boston taxicab driver. His story helped me understand that one doesn't have to follow a predetermined path to achieve something in life, that one can find one's way without knowing it at the outset. It gave me hope that I, too, could find my way.

From then on, I went to see Professor Sally in his office three or four times each week. I would show up at his office hours even when I didn't have any questions. I was simply drawn to the man, the most interesting and caring professor I had at Chicago. Every time I showed up at his office, his Boston accent would boom, "Hey man, give it there!" with a large fist extended in my direction. We had a special relationship, which continued until his death in 2013. I miss him deeply.

He suggested that I give graduate school a try. "After all," he said, "the mathematics one learns in college barely resembles the stuff that research mathematicians do. It might surprise you, and you might really like the stuff." Sally's confidence in me somehow gave me confidence in myself. He made phone calls on my behalf, and on the strength of his belief in my ability, he single-handedly got me into some of the top doctoral programs in mathematics. I accepted an offer to attend UCLA on scholarship. It would be a fresh start, in a much more forgiving climate.

It wasn't just the climate. There was Erika. We had been virtually inseparable for two years, and we knew that we would spend the rest of our lives together. She would need a second bachelor's degree toward becoming a nurse midwife, and UCLA had a strong nursing program. Erika had one more year at Chicago to complete her degree, and we wistfully agreed that spending a year apart was in our long-term interest.

Chapter 25

∼

GROWING PAINS

Los Angeles (1989–1991)

Ramanujan's story had offered me hope in 1984, when I was a depressed tenth-grader, that I could find my own path in life, and so I dropped out of high school and left home. My path was still crooked and uncertain, but on it, I had found emotional and intellectual support, first from my brother Santa in Montreal, and then in Chicago from college friends, fraternity brothers, my cycling mentor Tom Kauffman, and professors such as Paul Sally. Then in 1988, just as I was nearing the end of a half-hearted math major at Chicago, Ramanujan helped me a second time. The television documentary about his life reinvigorated my feeling of hope, and it inspired me to work hard my senior year. The documentary had caught me off guard, and it knocked some sense into me. I focused on mathematics my senior year and impressed Professor Sally enough that he helped get me into UCLA.

Of course, I was no Ramanujan. I wasn't a genius, and whereas he, as his mother's "little lord," grew up perhaps being told he could do no wrong, I as a child could do no right. But in spite of everything, I had earned a bachelor's degree from a first-rate college and had been accepted into a major graduate program in mathematics with a fellowship. And I was loved by the woman I loved. If she saw something in me, I couldn't be completely worthless, could I?

Yet I was plagued with doubt. It was clear to me that I was nothing but an impostor. As the son of a famous math professor, I could see that for success, you

© Springer International Publishing Switzerland 2016
K. Ono, A.D. Aczel, *My Search for Ramanujan*, DOI 10.1007/978-3-319-25568-2_25

had to work really hard, spending all day, every day, scribbling on yellow pads. I didn't have a prayer. I didn't have the necessary talent, and I didn't have the necessary devotion. The voices in my head confirmed the hopelessness of my case:

Ken-chan, your record at Chicago not earn you place at UCLA. You pull wool over Sally's eyes and he think you good enough to be real mathematician. UCLA foolish to believe him. You not good enough. You going to fail.

But voices or no voices, qualified or not, UCLA had offered me a teaching assistantship and a scholarship, and I had accepted and was on my way. I suppose I held out hope that I could turn things around in grad school. In any case, I could try. I had to do something, and this was the one open door. And who knows? I might somehow fulfill my dream of obtaining parental approval. In the end, I went to UCLA because I had nowhere else to go.

I arrived in Westwood Village in Los Angeles in August 1989, and I moved into a studio apartment with fellow UCLA math graduate student and UChicago alumnus Brad Wilson. Our apartment was three blocks from Pauley Pavilion, the famous basketball stadium where John Wooden, the "Wizard of Westwood," won ten NCAA basketball championships coaching UCLA with players like Lew Alcindor, who would later be known as Kareem Abdul-Jabbar, and Bill Walton. Westwood is sandwiched between Beverly Hills, home to Hollywood celebrities, and idyllic Santa Monica, home to its well-known namesake pier and beach. UCLA, a world-class university, seemed strangely out of place in this southern California paradise. I knew that I was in trouble the moment I stepped out of the cab in front of our apartment. How in the world was I going to earn a doctorate in this warm, sunny place covered with palm trees?

Despite our choice location, our apartment was a dump. We shared a gloomy room with a bathroom and a closet of a kitchenette. The rent was an astounding $628 per month, an enormous price for such a rathole. I would be spending over half my stipend of $7500 just on rent. And the exorbitant rents in our building were clearly not being invested in maintenance. The pipes in the ceiling above our bathroom leaked, and once a pipe burst, making me the unfortunate recipient of an unexpected cold shower.

I spent the weeks before the start of classes enjoying my new environment. I went for long bike rides on the Pacific Coast Highway, which offered stunning views of cliffs, miles of beaches, and famous places like Venice Beach and Malibu. I went mountain biking in Will Rogers State Park, familiar to me as the set for the classic TV show *M*A*S*H*. I enjoyed eating out for almost every meal at places like Fatburger, Hurry Curry, In-N-Out, and the Reel Inn. Los Angeles had so much to offer, and I wanted to experience it all. I probably should have been thinking about the PhD program, but there were too many distractions to which I simply had to yield.

I expected graduate school to be challenging, much harder than college. This was, after all, full-time professional training. No more general education, fraternities, and extracurricular activities. But as to how challenging, I didn't have a clue. My ignorance didn't last long. At orientation, Professor S.Y. Cheng, the director of the graduate mathematics program, welcomed us with an ominous warning. He predicted that one-third of us would drop out or flunk out within two years. If we were an exceptionally good group, then half of us would finish a PhD. Looking around the room, I queerly felt like a foreigner among my classmates. It seemed as though half of the new PhD students were Asian. But they weren't Asian-American like me; they were Asian-Asian, imports from China and Japan, where they had earned their undergraduate degrees. Many of them could barely speak English. But it clearly wasn't those students that Professor Cheng was worried about. He seemed to be speaking primarily to the home-grown students like me, and he sounded just like a tiger parent.

Along with most of my fellow first-year classmates, I enrolled in graduate courses in abstract algebra, real and complex analysis, and geometry and topology. We were expected to pass four qualifying exams within two years, and most of us would take exams on those topics. As a UChicago graduate, I was confident that I was adequately prepared. Alas, I couldn't have been more wrong. We at once plunged into these subjects at a much deeper level than what the introductory courses I had taken in college had prepared me for. To my surprise, the Asian graduate students seemed somehow to master the material with ease, despite their poor grasp of English.

I was in trouble, and I quickly lost confidence. My voices, in that familiar slow, accented drawl, spoke to me often:

Ken-chan, of course classes hard, too hard for you. What you expect? You don't belong here. All that time you spend on bike, these students prepare for graduate school. You not good enough to be mathematician.

I didn't realize that I wasn't alone. I soon found out that many of my domestic classmates were struggling as well. Some of them decided to pool their intellectual resources and form study groups, while I stupidly struggled almost entirely on my own. I also didn't realize that many of the foreign students had already earned graduate degrees abroad before coming to UCLA. They had seen this stuff already. It was not a level playing field.

I was alone with my anxieties, and Erika was far away. She was in her last year of college, enjoying a semester abroad studying French literature in Paris, living next to the beautiful Luxembourg Gardens. I wrote her dozens of letters in which I complained about my predicament. I told her that I was a fraud, an impostor, that I had somehow fooled Professor Sally into thinking that I had talent and potential, the right stuff to become a mathematician, that I was in grad school by a fluke, that I was in way over my head and sinking fast.

In the spring of 1990, after only one semester at UCLA, I began to plot another escape. I was afraid of the qualifying exams. Many students flunked those exams. And for those who passed, what were their prospects? Everyone knew that almost all the recent PhDs were struggling to get jobs. The most I had to hope for was in a few years' time to be one of those stressed-out students about to defend their dissertations while waiting to hear about their hundred-plus job applications, knowing full well that many of them wouldn't receive any offers. If the best graduate students at UCLA, those who actually finished a thesis, were having trouble getting jobs, then what chance did I have?

But that was all in the distant future. Right now, I was petrified of the exams, and writing a thesis wasn't even on my radar. Anyhow, it seemed like a mission impossible, like climbing Mount Everest without the aid of bottled oxygen, a feat reserved for those with superhuman abilities. My father had taught me that every thesis requires a solution to an unsolved problem, a question that others had failed to crack. How in the world was I, a mere mortal, going to do that? I finally managed to convince myself that I didn't need to run away just yet. I would take the qualifying exams, and in the likely event that I failed them, I would then leave UCLA with a consolation prize, a master's degree. It wouldn't be a triumphant outcome, but not a total failure either, though my parents would probably think so.

To prepare for the abstract algebra and analysis qualifying exams, I studied two to three hours every day for months. I pored over old exams, memorizing the problems and their solutions. It seemed like a good idea at the time. Abstract algebra was my area of interest, and so it was critical for me to pass that exam, for otherwise, I would be mercilessly exposed as the fraud I was afraid I was.

I took both exams, and boy, was I nervous! I found them to be quite difficult, but then they were supposed to be challenging, and we all understood that a score of sixty percent would be a passing grade. I was not able to solve all the problems, but I was confident that I had scored enough points to pass them both. Several weeks of slow-drip torture passed before we learned our fate. Time slowed to a crawl, and then it slowed some more. I was reminded of my days as a preschooler, when my mother would take me with her to the local art museum, where I waited, waited, waited while she painstakingly copied one of the paintings.

All of us who had taken the qualifiers were waiting for our scores, which we knew would come by way of a letter placed in our departmental mailbox. I did not exactly camp out in front of my mailbox, but for several weeks, whenever I was at the department, I would check the mailboxes every hour. I felt as though my entire future were riding on whether my score would be below sixty percent or above. The day and hour finally arrived when two envelopes appeared in my mailbox. This wasn't like college acceptance letters, with a thick letter for acceptance and a thin one for rejection. If I wanted to know whether I had passed or failed, I was going to have to open the envelopes.

The first letter I opened was for the analysis exam. I had passed! What a relief! The sun was shining, all was right with the world. It was going to be a great day, the day I overcame two important hurdles on my way to a PhD. I had been more nervous about the analysis than the algebra, so I opened the second envelope prepared to whoop with joy. But I had failed. I had failed the algebra exam, the one that I absolutely needed to pass. "Ken-chan, you a fraud." I was in shock.

I was also seized by a rational sense of disbelief. I couldn't have done so poorly. I ran to the graduate office and requested a photocopy of my algebra exam. I stormed off home with it to perform an autopsy.

Immediately on my arrival, I added up my points. I added them up again, and the sum again refused to rise above the magical sixty percent. How could I have screwed up so badly? I went through the test, question by question. What was this? Here was a problem that I had certainly solved correctly, but I hadn't been given any credit for it. Give me those points, and I pass. I was exultant, but I required corroboration.

In a fit of self-righteousness, I felt that I had to locate someone in a position of authority immediately to countermand my failure. I would explain my solution, and I would be reinstated among the elect. My nightmare would be over. Professor Elman, who had taught the first-year graduate course in abstract algebra, was the person I needed to talk to. I looked up his number in the phone book, and I called him at home. I read him my solution, and he agreed that it was indeed correct and that I should have received credit for it. But then he added, to my horror and disbelief, that a mere passing grade in my chosen field was not good enough. That I had passed the exam did not mean that I had performed adequately. If I planned to write a thesis in abstract algebra, I should have aced the exam. Squeaking by with a pass would not augur well for my future success.

I hung up the phone completely deflated. I sat on the couch in our dump of a room stunned, frozen in fear and chagrin. I had passed both exams. I had not failed. I had been successful. A pass was a pass, wasn't it? Why did Professor Elman not see it that way?

I was angry. I had some important business to attend to, and I wanted it to be accompanied by the triumph of the successful completion of my qualifying exams. I did not need to be told, yet again, that I was not good enough. So I left Westwood in anger. My classmate Bruce Abe and I rented a car and drove the nearly thirteen hundred miles to our destination. I had a lot on my mind on the drive as Elman's words reverberated in a continuous loop. At times, they assumed a familiar accented drawl: "Ken-chan, passing grade not good enough. You need do much, much better." We crossed the searing Nevada desert, and after a brief stop in Salt Lake City to see Bruce's mother, we arrived in Missoula, Montana.

I had come to Missoula to get married. The long drive had been good for me. I was able to decompress and get that infinite tape loop about my failure in abstract algebra out of my system. By the time we got to Erika's parents' house, my sense of doom was long gone. I was thinking about my new life with Erika.

Jan and Robin Anderson, Erika's parents, are the friendliest people I have ever known. They are constantly inviting people to their home, be it relatives, friends, or even friends of friends. And so there was never any question as to where the wedding ceremony would be held. It had to be at their cedar-clad mountain home, which overlooks the Missoula Valley with the majestic Bitterroot Mountains on the horizon.

Erika and I were married on June 23, 1990, on the front lawn of her parents' home. It was a beautiful day shared with many friends and family members, including my parents and both my brothers, who had flown out for the occasion. Amazingly, my uncle and aunt Yoshitaka and Natsue Nomachi arrived by plane from Tokyo. Ron, my Pepsi-Miyata teammate, was one of my groomsmen. Erika's sister, Holly, was our maid of honor, and Santa was my best man. Momoro played a lovely rendition of Pachelbel's *Canon* for the opening processional. We recited our vows and then enjoyed grilled salmon, wonderful music performed by our families, and dancing on a makeshift platform well into the night.

The wedding festivities provided a much-needed escape from my qualifying exam nightmare. June 23 would now be a special day for the rest of our lives. We had no idea that it would soon be important to us for another reason. One of the most important events in my life would take place three years later to the day. It's not what you think.

Newlyweds (photo by Jan
Anderson)

Erika and I returned to Westwood as a married couple, and we moved into a small studio apartment a few blocks from the one I had shared with Brad. We bought our first car, a dark blue Hyundai Excel hatchback. It was more go-kart than car. We bought the cheapest model, the one where none of the buttons on the control panel actually did anything. Erika found work in a Beverly Hills office, where her boss was Kelly Stone, the sister of Sharon Stone, the sultry Hollywood actress who was about to star with Michael Douglas in the psycho-thriller *Basic Instinct*. Thanks to Kelly, Erika and I would end up meeting all sorts of Hollywood figures. In fact, one of those Hollywood connections would soon get us our second apartment. We had gotten a tip on a rent-controlled apartment in Santa Monica from an actress who had a minor role in the TV soap opera *General Hospital*.

That apartment was a cute little bungalow on 16th Street, one block from Montana Avenue. We loved the location. I went for runs to the beach on pictur-esque San Vincente Boulevard, and we made frequent trips to the Third Street Promenade, where it was not unusual to bump into a celebrity such as Jane Fonda, Julia Roberts, Arnold Schwarzenegger, O.J. Simpson, or Ted Turner, among too many others to mention.

Early married life is a sweet time of transition, and for me, part of the sweetness was putting things in perspective and adopting a longer view of the future. As I reflected on my first year at UCLA, I came to realize that Professor Elman had been right about my relationship to abstract algebra. He was very wise, and he understood what was best for me. If I was going to train as an algebraist, it wasn't enough merely to pass the abstract algebra qualifying exam. I needed to ace it. I had to prove to myself and my professors that I was a worthy PhD candidate, one whom a potential advisor would be eager to take on. Professor Elman had thrown down the gauntlet, and I eagerly took it up with a vow not merely to pass abstract algebra, but to master it.

I realized that my earlier approach of reviewing and memorizing old exams was inadequate. It was more important to understand the underlying theorems, structures, and techniques than to know how to solve a small subset of all the possible problems that might be thrown at me. This time, I prepared with a completely different attitude. My goal was to try to understand the material well enough to teach the course. It is said, after all, that one never really learns a subject until one has taught it. And if I ultimately finished my doctorate and obtained a university position, then I would be expected to teach such a course. I approached my studies as though I were training for an important race. I systematically studied every theorem, how it was proved and to what kinds of problems it could be applied. The next time the exam was offered, I took it with confidence. I answered all the questions, completing the exam in half the allotted time. I knew that I had throttled it. I was so happy that I left the exam room hooting and fist pumping for joy.

Soon after I passed my qualifying exams, I was awarded a master's degree. This meant that I was due a small raise as a teaching assistant. But I had a different idea. I applied for a full-time teaching position at Woodbury University, a small school in Burbank. Amazingly, they offered me a job teaching three college algebra courses a semester at a salary of $25,000. That was a huge raise over my teaching assistantship. Moreover, I could teach the early morning classes, the ones nobody else wanted to teach. By taking the job, I would earn a much higher salary, obtain valuable teaching experience, and still be able to work toward my doctorate by attending classes and seminars in the afternoon at UCLA. Some of my professors and classmates questioned my decision, arguing that it would take time away from my studies. As it turned out, it was one of the best decisions I ever made, for at Woodbury, I learned that I love to teach.

Part IV
Finding My Way

Chapter 26

MY TEACHER

I finally passed my remaining qualifying exams, and on paper, I had earned the right to advance to candidacy in the UCLA doctoral program. This may have been true on paper, but in reality, I had no idea what I was supposed to do next. I was poised to drift through the program or else drift right out of it. I needed direction.

In the spring of 1991, I took an algebraic number theory course from Professor Basil Gordon. Gordon loved the material, and the students in the class could sense his deep devotion to the subject. While Herstein's lectures in his number theory class at UChicago were intimidating tirades, Gordon's lectures were inspirational sermons. It was like being at a poetry reading. For him, a theorem was not just some odd mathematical fact. It was a work of art whose aesthetic qualities could be described, as could its place in the ongoing intellectual dialogue of mathematics and the questions it raised for further research. Gordon would sometimes compare a theorem to a famous work of art or classic poem. It was not unusual for him to juxtapose the majesty of a theorem of Gauss with the breathtaking beauty of a Michelangelo sculpture. I soon understood that Gordon's relationship with mathematics was unusual. He viewed himself as an artist whose medium happened to be mathematics. It was clear that he thought about mathematics in a way that was very different from my view, which had always involved performance on exams and the memorization of formulas and proofs. I wanted to know more, and I didn't have to wait long to get my chance.

It was several weeks into the course, during a lecture about ideal class groups, a subject developed by Gauss a century and a half earlier. Gordon was just finishing the proof a theorem about prime-order torsion elements in these groups using a method introduced decades earlier by MIT mathematician Nesmith C. Ankeny and Penn State professor Sarvadaman Chowla. As I listened, it began

© Springer International Publishing Switzerland 2016
K. Ono, A.D. Aczel, *My Search for Ramanujan*, DOI 10.1007/978-3-319-25568-2_26

to dawn on me that there was a much more conceptual proof that made use of elliptic curves. After Gordon completed the proof, I raised my hand and offered my alternative proof, which made use of geometric ideas of Mordell and Weil. Gordon's response was to ask my classmates to applaud my proof, and he invited me to his office after class.

I nervously made my way to his office, worried that he would scold me for my presumptuousness. Had he been mocking me when he asked the class to applaud?

Gordon's office seemed strangely out of place at UCLA. It could have been the office of an Oxford don or one from the Hogwarts School in the *Harry Potter* novels. The walls were lined with beautiful barrister bookcases, their contents beckoning from behind hinged glass doors. An enormous ornate Persian carpet graced the floor, concealing the weathered 1960s-era floor tiles. The desk over-flowed with papers and letters, nearly enveloping a set of antique gold pens.

I was in the presence of a gentleman, someone whom I could imagine sitting by the fireplace with G.H. Hardy in the Reading Room at Trinity College enjoying a cup of tea. Gordon and Hardy? Come to think of it, Hardy had also compared mathematics to art, music, and poetry; perhaps I was onto something. Gordon's manner evoked images of a different time and place, perhaps a nineteenth-century English manor. Our discussion was brief. Although the proof I had offered in class was not a new result, he had been impressed with my insight. He had been following my career at UCLA, and he told me that he would be honored to be my doctoral advisor. He was thinking about retirement, and he wanted me as his final PhD student. Although I was surprised and puzzled by his offer, I accepted on the spot. That meeting with Gordon marked my birth as a mathematician.

Basil Gordon was indeed a gentleman and a scholar, a polymath who was a direct descendant of the Gordon family of British distillers, producers of Gordon's gin. He was the step-grandson of the famous American general George Barnett, who served as the major general commandant of the Marine Corps during World War I. I was pleased to learn that we had both grown up in Baltimore. He had attended Baltimore Polytechnic Institute and received his master's degree in mathematics from Johns Hopkins University in 1953. He earned his doctorate in mathematics and physics from Caltech in 1956 working under the mathematician Tom Apostol and the iconic physicist Richard Feynman. Gordon was drafted into the U.S. Army, where he worked with rocket scientist Wernher von Braun. He was part of the team that worked out the path of the satellite Explorer I so precisely that it remained in orbit for a full dozen years after its launch in 1958. Gordon joined the UCLA faculty in 1959.

With Basil Gordon in 1992 (photo by Keith Kendig)

Gordon and I developed an unusual routine. Instead of weekly sessions in his office at UCLA, we met at his home in Santa Monica. Gordon lived in an elegant house on Palisades Avenue, a quiet street a few blocks from Santa Monica Beach, about a mile from our bungalow on 16th Street. That was convenient for me, and since our meetings would last for hours, working with him at his home on Saturdays meant that we could work without interruption. That home was more museum than residence. There were hundreds, perhaps thousands, of books, a grand piano, and antique crystal. Original works of art adorned the walls.

We rarely began our meetings by diving straightaway into the mathematics. Instead, Gordon might begin by playing a Chopin nocturne on the piano. Sometimes, he would recite poetry from memory. Gordon had a photographic memory and could effortlessly quote reams of literature. I recall him intoning the first few pages of Melville's *Moby Dick*. He'd close his eyes, entering a trance-like state, and then begin:

Call me Ishmael. Some years ago—never mind how long precisely—having little or no money in my purse, and nothing particular to interest me on shore, I thought I would sail about a little and see the watery part of the world.

To leave the safe familiarity of the shore and sail off into unknown territory, that is what it is like to do mathematics. Gordon was constantly reminding me that our mathematical research, as difficult and as confusing as it can be, is an art form, an exploration, an adventure, something to be appreciated, something to be lived. How could we possibly prove a good theorem if we viewed mathematics as a chore? We weren't hanging sheetrock, we were creating a masterpiece, cultivated over weeks, months, even years of deep thought and imagination. And so it was music and poetry that set the tone before we began scribbling figures and equations on our yellow pads.

I learned a great deal from Gordon on those Saturday afternoons. We would spend hours huddled in his den, struggling with difficult concepts, trying to break an impasse and find a way to bridge a logical gap in an argument. From time to time, on rare occasions made the sweeter for their rarity, we were rewarded with a breakthrough, an elegant argument, a watertight proof. Those moments of revelation were so awesomely gratifying that we quickly forgot the doubt and despair that can creep into the soul when one has lost one's way.

Then we might go out for lunch at the corner Italian bistro, followed by a long walk to the beach. We must have looked an odd couple—an Asian-American young man in his early twenties with a mullet haircut and neon clothes strolling slowly with a sixty-year-old gentleman, nose painted in zinc oxide, dressed in khaki pants, polo shirt, and low-cut white canvas sneakers. Perhaps not so odd; this was, after all, Southern California. But anyone catching snippets of our conversation would have been baffled by our passionate outpourings about continued fractions, modular forms, and Galois representations. Of course we were passionate. We were talking about great mathematical works of art created by the likes of Gauss, Euler, Galois, Serre, Shimura, Taniyama, Weil, and, of course, Ramanujan.

From Basil Gordon I learned what it means to "do mathematics." When I was a child, I understood as a child, and I thought that math was only about manipulating numbers and "solving for x." In college, I thought mathematics was about memorizing theorems and proofs, mastering techniques for carrying out difficult calculations, and solving textbook problems.

Gordon taught me that "doing mathematics" begins with a state of mind that allows you to travel to a place deep inside the subconscious to open body, mind, and spirit to the contemplation of a mathematical idea. Doing mathematics is a mental voyage in which clarity of thought and openness to insight make it possible to see the deeper beauty of a mathematical structure, to enter a world where triumph over a problem depends less on conscious effort than on confidence, creativity, determination, and intellectual rigor.

Although I had never practiced such meditative techniques with mathematics, it sounded just like what I had been doing for years on my bicycle, when, for example, on my solo training rides, I was racing the likes of Eddy Merckx up the imagined slopes of Mont Ventoux. The state of complete concentration that I achieved helped me find the strength to pedal a higher gear, propelling me a bit faster than I thought I could manage. Gordon taught me that a similar practice could be instrumental in discovering and proving difficult theorems that might otherwise be off limits.

When I do math now, I enter a trancelike state in which I travel to a special place to visit my old friends, objects called tau, $p(n)$, and $f(q)$. If you looked at me while I was working at home, you might think that I had dozed off. I also do much of my best work on long solo runs and rides on my mountain bike. Those activities make me more alert and mentally productive, both while I am exerting myself physically and in the hours afterward.

I would later find my spiritual self. Like many scientists, I would come to believe that the unimaginable complexity and symmetry exhibited by numbers can be nothing less than evidence of their divine origin. The deep trancelike states and the mental freedom that I achieve in running and cycling are spiritual. They give me an openness of mind and soul that allows mathematical secrets to be revealed. Perhaps it is something like what Ramanujan experienced when he received visions from the goddess Namagiri.

Gordon encouraged me to study the theory of modular forms, a subject that has its origins in much older works of Euler and Jacobi. He steered me in the direction of the more modern treatment of the subject as developed by Pierre Deligne, the Belgian mathematician who proved a conjecture of Weil that in turn solved a problem of Ramanujan. He also had me study papers by the French mathematician Jean-Pierre Serre, who together with Weil inspired my father and the other young Japanese mathematicians at the 1955 Tokyo–Nikko conference.

I learned that the work of Deligne and Serre on a subject called *modular Galois representations* was somehow rooted in the work of Ramanujan. I had not thought about Ramanujan for two years, and so it came as a surprise to learn that his work played a role in the development of cutting-edge mathematics. Serre wrote a lovely article in 1973 in which he explained how the English mathematician Swinnerton-Dyer had recast some old results of Ramanujan, who had obtained them by means of the masterful manipulation of power series, into the geometric framework of infinite Galois theory that Deligne was developing.

Ramanujan's formulas, known as *congruences for tau*, exemplified and somehow anticipated the deep conjectures of Serre that Deligne was proving at the time, work that contributed to the Fields Medal that Deligne was awarded in 1978. (Serre was awarded the Fields Medal in 1954. Serre and Deligne are two of only four mathematicians to have been awarded the Fields Medal, the Wolf Prize, and the Abel Prize.) Deligne's research proved Ramanujan's congruences as well as the far-reaching Weil conjectures. Ramanujan's seemingly old-fashioned formulas seemed deeply and strangely intertwined with the ultra-modern theories developed by those great mathematicians.

My dissertation research was forging another connection to Ramanujan. It would not be long before that connection would be transformed into a personal search for Ramanujan the mathematician. But I still had to go a very long distance out of my way before that would happen.

During my second year at UCLA, Robert Kanigel published his book *The Man Who Knew Infinity: A Life of the Genius Ramanujan.* I bought the book as soon as it came out and almost devoured it in a single sitting. It offered vivid descriptions of exotic sites in south India, and it filled in many details about Ramanujan's life, about which I had actually known quite little. Ramanujan suddenly became for me more than a source of magical formulas. He became a mystical figure whose life seemed to contradict every stereotype I had of mathematicians. Kanigel's description of Ramanujan's life made it seem like something more out of the Arabian Nights than the history of twentieth-century mathematics. It was a most improbable tale.

After finishing the book, I felt that I would like one day to make a trip to India to pay homage to Ramanujan. Later, that desire would grow to the point that I would feel compelled to make a pilgrimage to search for Ramanujan himself, the mathematician and the man. It would become my calling.

Kanigel's book included expert commentary from mathematicians George Andrews, of Penn State, and Bruce Berndt, of the University of Illinois. Berndt was devoting his career to working out Ramanujan's unproven claims, systematically working through his writings, making sense of the Indian genius's assertions and supplying proofs wherever they were lacking. This task would take him decades to complete, and he did not work alone. He enlisted the help of many mathematicians, mostly doctoral students and newly minted PhDs. Berndt was a mathematical guru, who helped many young mathematicians at early stages of their careers. Although I didn't know it at the time, those two men, Andrews and Berndt, would soon play important roles in my own life.

George Andrews in 2008
(photo courtesy of the
American Mathematical
Society)

Bruce Berndt with
Ramanujan's slate in the
1980s (photo courtesy
of Bruce Berndt)

For the next two years, I worked on my dissertation and taught classes at Woodbury. I filled my need for physical activity as a slow member of the Santa Monica Track Club, which boasted Olympic gold-medalist Carl Lewis among its lightning fast members. Erika and I enjoyed life as a young married couple in picturesque Santa Monica. It should have been a halcyon time, but in the depths of my psyche lurked still the parental voices that ate away at my self-esteem like the eagle gnawing eternally at Prometheus's liver. I survived thanks to Erika's and Gordon's nurturing. Without either of them, I would have surely dropped out of UCLA, perhaps creating a second black hole of memories.

I didn't expect that my thesis would be of interest to many mathematicians, but that didn't bother me. A thesis is supposed to be the first step in the life of a professional mathematician, not the final magnum opus. Gordon had taught me to love mathematics for its own sake, and that discovery sustained me. My goal was to finish the dissertation, and Gordon assured me that I was well on my way to success. Strangely, I believed him. We were having a wonderful time proving theorems, and proving them for their own sake.

Chapter 27

HITTING BOTTOM

Montana (1992)

*B*asil Gordon had awakened me. I had been dangerously adrift at UCLA, and his mentoring coaxed me back into the life I was meant to lead. Like Ramanujan, I had developed an addiction for mathematics. I was in love with mathematical beauty. I had a passion for doing mathematics. Yet I had no idea how things might turn out for me professionally. Desire does not always lead to fulfillment. Would my theorems be good enough for anyone to care about? Might I have a thesis in me but not much else?

Although I was pleased with Gordon's assessment of my progress, I didn't entertain any thoughts of a top-flight research career. The voices in my head told me that I had no chance. So I set my sights lower. My wish was to land a position at the University of Montana in Missoula, Erika's hometown. To secure a teaching position at UM would have been a dream come true. We would have bought a house near campus and set down roots in that lovely college town, with Erika's family nearby.

In the spring of 1992, I learned about a conference in Missoula, one that I thought might offer the opportunity of a lifetime. The Pacific Northwest Sectional Meeting of the Mathematical Association of America (MAA) would be held at the University of Montana in June. I was ecstatic to learn that the distinguished plenary lecturer at the meeting would be Bruce Berndt, the celebrated Ramanujan expert about whom I had just read so much in *The Man Who Knew Infinity*. He was also masterfully training many young mathematicians. If I could meet him, then maybe he could help me, just as André Weil had helped my father in 1955 at the Tokyo–Nikko conference.

© Springer International Publishing Switzerland 2016
K. Ono, A.D. Aczel, *My Search for Ramanujan*, DOI 10.1007/978-3-319-25568-2_27

I couldn't believe my luck. History seemed poised to repeat itself. The planets were in alignment, and good fortune was headed my way. I made plans to take full advantage of the opportunities. I would attend the conference and there make a name for myself, impressing both Berndt and the math faculty at the University of Montana.

Another and much larger meeting was scheduled four weeks after the Missoula meeting: the Rademacher Centenary Conference at Penn State. George Andrews, the celebrated Ramanujan scholar whom I first saw on TV at my frat house in 1988, was one of conference organizers. He had earned his PhD under Hans Rademacher, a mathematician whose fame was rooted in work that perfected one of Ramanujan's most important theorems. Andrews was one of the organizers of this conference to honor the centenary of his advisor's birth. Both Andrews and Berndt would be there, and Gordon also was among the invited speakers.

Together with Gordon, I hatched a plan: I would submit abstracts to both conferences proposing short contributed talks. I would also write some of the UM math faculty and offer to give a seminar before the MAA meeting in Missoula. I had a place to stay, the comfort of Erika's childhood home. Gordon believed in me, and he assured me that I was ready for both meetings.

I asked Doug Bowman, one of Gordon's other PhD students, to accompany me to Missoula. The promise of a road trip to Montana convinced Doug to attend the meeting with me.

Like Ramanujan, Doug was a self-trained mathematician who recorded his findings in notebooks. He studied a subject called q-series, which happened to be one of Ramanujan's areas of expertise. Doug had been publishing papers for years, and as a result, he was attending UCLA on a prestigious graduate fellowship awarded by the National Science Foundation.

Doug and I submitted our abstracts to the MAA meeting, and we were delighted when they were both accepted. Through a family friend, I was introduced by email to Professor George McRae, one of the senior professors in the UM Department of Mathematics. He invited me to give a departmental seminar before the scheduled meeting. More precisely, he kindly acceded to my offer to present a seminar. My plan seemed to be working. I would get to meet the famous Professor Berndt, and I would have an opportunity to impress the UM faculty. All I had to do was deliver.

I prepared both Montana lectures with care—rehearsing them several times. Several days before the conference, Doug and I left Los Angeles in my go-kart of a car. Since we had thirteen hundred miles to drive, it made sense to enjoy some well-deserved R and R before the important seminar and conference. Erika didn't come along; she couldn't get time off from work.

Doug and I enjoyed our road trip from Los Angeles to Missoula. We drove through the searing desert in my car, which of course had no air conditioning, listening to Liz Phair, New Order, and other alternative rock bands. We spent a day in Las Vegas, walking the "Strip" and playing the slots at Circus Circus. We also made a side trip to Bryce Canyon National Park to marvel at the enormous amphitheaters of inverted red rock spires.

We were both excited to meet Bruce Berndt, and I was hopeful that the trip would lead ultimately to a job for me at UM. We couldn't stop talking about what we hoped would happen in Missoula.

Instead, disaster struck for me, and it struck twice. My seminar at UM was an unmitigated catastrophe. I had devoted my fifty-minute lecture to describing my research on Galois representations, a topic that I had been thinking about for over two years. But nobody at UM, as I should have known, knew anything about those objects. Like most research mathematicians, I was working in a sub-specialty that few mathematicians working in other fields would understand. I suppose I had figured that if I, a mere third-year graduate student, understood the stuff, then all those senior professors should understand it too. I should have been talking in broad generalities about my subject rather than the minute details that only experts in the field would care about.

But I wanted to make an impression, and I figured I could do that by presenting the theorems that I had proved. So I began my lecture with a short and per-functory introduction, so as to leave ample time for the exposition of my results.

Imagine that you are a rocket engineer invited to lecture about rocket design to a room of engineers who are not, if you will pardon the expression, rocket scientists. You begin your lecture with a slide of a slingshot, followed by a brief history of ballistics and the early history of rocketry. But after five minutes of this, you jump into your minuscule area of expertise and spend the next forty-five minutes explaining in painful detail the calculations behind the latest improvements to oxidizer design. I was the mathematical version of that mis-guided engineer.

The rest of my talk was confusing and frustrating to the audience. And confused and frustrated audiences tend to stop listening after a few minutes. I doubt that anyone in the room paid any attention to my lecture after the first five or ten minutes. They were lost, and it was my fault for not taking the time to motivate my subject. In fact, my talk should have been mostly motivational with a brief mention of my results at the end. My theorems were the equivalent of oxidizer design. They did not merit such a full-court press in a presentation to a general audience. But I had wanted to strut my stuff.

After my lecture, McRae kindly offered me a bit of advice. I had not thought carefully about my audience. McRae was a gentleman, a warm man whom I continue to revere. Yet his soothing words were quickly forgotten when another professor approached me in the presence of many others in the department lounge. Sputtering with rage, he reproached me for wasting his time, fuming, "I have to say, I am a world leader in my field, and you aren't. Have something to say before you decide to talk." He then turned on his heels and stormed off, shaking his head in disgust without giving me an opportunity to apologize. What could I have said? As a twenty-four-year-old graduate student, I was an inexperienced lecturer, and the UM seminar was my first such presentation. I was devastated.

McRae overheard a bit of the conversation, and he tried to mollify me. He whispered that this professor had a reputation as an imperious grouch, who, far from being a world leader, actually had difficulty getting his papers accepted for publication.

But his words didn't help. I was beyond help. I hadn't given McRae, or anyone else, any reason to believe that I had proven much of anything. And in any case, my primary goal in giving the seminar should have been to prove that I was knowledgeable about my subject and could present my work to a diverse audience of mathematicians. I had done everything wrong, and I was distraught. I had hoped to impress the faculty, and instead I had fallen flat on my face. It didn't matter whether my theorems were worth anything. I had given such a poor lecture that nobody was able to evaluate what I had done, not even the context in which I had done it. And I had pissed off a senior professor to boot.

Talking to Erika on the phone that night, I learned that she had attended high school with the angry professor's son. And so a person who might well have been an advocate for me when I applied for a position at UM now despised me for wasting his time. The voices in my head had a new companion, another real live person to join my parents and the junior professor at Chicago. And this voice belonged to a mathematics professor at UM who even knew Erika's family. It was humiliating. I worried that news of my poor performance would reach Erika's parents. What would they think? I had arrived in Missoula hoping to make friends and influence people, and instead I had made an enemy and alienated everyone.

Disaster struck again a few days later at the MAA conference. After Berndt's breathtaking plenary lecture on Ramanujan, Doug and I gave our contributed talks. I had carefully prepared a set of overhead projector transparencies, and I felt unable to modify my talk based on what I had learned from my failure at the UM seminar. Going into the talk, I knew I was in trouble. I was a dead man walking.

Berndt attended both of our talks. He congratulated me on an interesting presentation, but I could tell that he was just being polite. On the other hand, Berndt was deeply impressed by Doug's talk, and at lunch, I overheard him invite Doug to apply for a tenure track position at the University of Illinois at Urbana-Champaign, one of the top mathematics departments in the country.

Doug had done me a favor by agreeing to accompany me to Montana in the first place, and he had offered lots of sage advice on our long drive from Los Angeles to Missoula. And then at the meeting, which was so important to me, it was he who had ended up as the young star. He had been the one to impress Berndt, and I was at most a mere afterthought.

Doug certainly deserved the recognition he had received. But his stunning success and my abject failures were more than I could take. I told myself that I was happy for him, but actually, I was furious at him. I felt betrayed. I was the one who needed a success, and he had stepped in and carried off the prize. "You, too, Brutus," I kept repeating, although neither Doug nor anyone else had stabbed me in the back. It was more like I had shot myself in the foot. Both feet, to be precise.

The voices in my head, the ones I had been struggling with for almost ten years, had been vindicated. I was no good, and despite my meticulous planning and preparation, I had failed, and my friend served as a barometer by which I could measure my inadequacy. Some are born great, some achieve greatness, says Shakespeare, but some, like me, are born to fail and achieve failure. My parents had always been critical of me, and now I worried that word of my disastrous performance would reach Erika's parents. I had come to Missoula with such high hopes. The result was worse than anything I could have imagined.

The last day of the conference included a social event at Salish-Kootenai College, in Polson, a one-hour drive from Missoula, where the lectures had been held. Doug and I had planned to drive together. But then Doug accepted an invitation to ride along with Berndt, who was clearly courting him. There wasn't room in the car for me, and since I had my own car, I decided to drive alone.

I like to drive with the radio on. But in the mountains of western Montana in 1992, my little go-kart couldn't tune in any stations. It was raining, and I had only the voices in my head to keep me company:

Ken-chan, you work hard, but you not good enough. Some professors kind to you, but you should not trust people who only have kind words. The critical professor is the one who speaks truth. Truth is that you wasting people's time.

I had spent months preparing for my seminar and the contributed talk. Erika and I had talked at length about our high hopes for the future. We believed that Berndt's presence and my seminar in her hometown were omens of imminent good fortune. We had even tempted fate by looking into the price of houses near campus. What was I going to tell her? How could I tell Erika that I had destroyed our chance for happiness? "Ken-chan, you spoil everything."

On the road to Polson, near a place called Ronan, there is a long straight section that stretches a mile or two downhill. From the top of the hill you can see oncoming traffic long before it reaches you. When I saw a logging truck approaching in the distance, a plan of escape began to unfold. I visualized in slow motion the truck smashing into my car, first crumpling the hood, followed by a beautiful spider-web pattern of shattered windshield exploding in my face in a violent shockwave of wind, glass, and rain. The truck came closer and closer, and when it was close enough for my purposes, I swerved across the yellow line.

I don't know what saved me. Perhaps it was the frantic blare of the trucker's horn. All I can recall is coming to a stop on the side of the road and sitting in the car in the pouring rain with the engine running, shaking and dripping with sweat. I thank God for saving my life.

I couldn't believe what I had almost done. Thwarted by the outcome of the disastrous meeting, I had lost all hope, and I was no longer myself. My actions seemed eerily to mirror Ramanujan's own suicide attempt when his nomination for a Trinity College fellowship was denied. I had never had suicidal thoughts before, and I was frightened by what I had almost done. It was an impulsive act that I will never fully understand.

The next day, Doug and I left Missoula for our long drive back to Los Angeles. He had impressed a world expert, and his career was about to take off. (Two years later, Doug accepted a position at Urbana-Champaign.) Although I was happy for him, or at least that's what I told myself, hearing him talk about his unexpected good fortune was maddening. I felt that I was on the verge of a nervous breakdown. I couldn't escape the sense that the voices in my head had triumphed, and they were predicting a bleak and shameful future. There would be no André Weil to discover me as he had discovered my father in 1955. If there was ever a confluence of events that was supposed to serve that purpose, this meeting had been it. "Ken-chan, you impostor. You sweet-talk Paul Sally and Basil Gordon. Now you see what you really worth."

I had hit bottom, and just when things had been looking up. Gordon had transformed me. I was now a budding mathematician who saw beauty in formulas and theories. I couldn't stop thinking about mathematics. I had become addicted to mathematics like Ramanujan. But the rejection of my work, when I was so passionate about it, was almost more than I could take. That it wasn't actually my work that been rejected, that I had simply given a couple of poorly conceived presentations, was something I was unable to see at the time. Unlike Ramanujan, who before being "discovered" by Hardy had been working all alone without recognition and had every reason to be depressed, I had a mentor who was carefully nurturing me. Nevertheless, in my depression I identified with Ramanujan, perhaps as a way of feeding my self-pity.

Chapter 28

⌢

A MIRACLE

Los Angeles (1993)

I didn't tell Erika very much about the seminar and the conference. I couldn't bring myself to admit to her that all our hopes had been dashed. I couldn't tell her how badly I had failed. I also dreaded my next meeting with Gordon. I had let him down. I had imagined a triumphal march down Palisades Avenue to Gordon's house to the sound of cheering crowds and popping champagne corks. In our time together, we had already had much to celebrate—mastering a difficult research paper, completing the proof of a theorem, the acceptance of a paper for publication. But there was nothing to celebrate now. I had pain to share, and I wanted to spare him, and myself. On the day of our usual meeting, I walked down Palisades Avenue accompanied by no sound but my beating heart. I reached the house and stepped up onto the porch, and then I froze. I must have stood for five full minutes staring at the heavy oak door before I could muster the courage to press the doorbell. I felt as if I had arrived at my own funeral. What was I going to say?

I began our meeting by giving Gordon a play-by-play account of the Missoula trip, and he listened attentively with his eyes closed, frowning and grimacing at the most painful moments. I could see that he was sharing my pain, visualizing the events as I retold them. I didn't tell him about my brush with death near Ronan. I couldn't.

After I finished relating most of the sordid details, he took a deep breath, and while staring off into the middle distance with his eyes wide open, he spoke slowly and deliberately: "If you can dream it, then you can do it." It would not have been unusual for Gordon to have recited an epic poem or lines from Shakespeare, but Disney? After a long pause—it must have been at least a minute—he said it again, this time in full: "If you can dream it, you can do it.

© Springer International Publishing Switzerland 2016
K. Ono, A.D. Aczel, *My Search for Ramanujan*, DOI 10.1007/978-3-319-25568-2_28

Always remember that this whole thing was started with a dream and a mouse."
After another long pause, Gordon began to tell me why he had asked me to be
his last PhD student.

He told me that he had felt that we were somehow destined to work as a team.
There had been signs. We were both raised in Baltimore, and we both were
identified as math prodigies at an early age. Although he had never met my
father in person, he had read his papers and books. He had studied my father's
important theorems on Tamagawa numbers and algebraic tori, and he had
followed his more recent work in algebraic number theory. He was therefore
delighted when I registered for his algebraic number theory course, and he was
thrilled by my obvious interest in the subject.

Gordon saw beauty everywhere, but it was more than beauty. There was
something spiritual in the things of this world, but also in the creations of the
mind, so that along with Tennyson's "The sun, the moon, the stars, the seas,
the hills and the plains," he felt that art, music, poetry, and mathematics were
also the vision of a higher power. Without a family of his own, he viewed his
PhD students as his children, and in me he had felt a special bond that had
begun from his longtime admiration of my father. There could be no more
beautiful way to end his career than to advise me, poetically helping to extend
my father's legacy.

He had realized early on that I needed not only mathematical advising, but
emotional support as well, and he had felt that he could provide both. He saw
that I, like many other graduate students, viewed the doctoral degree as the
single goal of graduate school, with coursework, qualifying exams, and the
dissertation a series of hurdles to be overcome. I had been seeking a credential
for the credential's sake, a ticket that would allow me to move on to the next
credential. Now after the many months we had spent working together, he
explained that I had been transformed, that mathematics was no longer for me a
means to an end but an end in itself. I had matured into a scientist. I was a mathematician. It was the creation of beautiful mathematics that was the true goal.
Then he thanked me for sharing my transformation with him. I couldn't believe
it; he was thanking me when I was the one who should have been thanking him.

Instead of dwelling on the pain of Missoula, he had redirected my thoughts.
I looked inside myself. He was right. I had become a different person. I was
enjoying mathematics for its own sake. Under Gordon's tutelage, the walks on
the Santa Monica boardwalk, the poetry, the Chopin nocturnes, and the mathematics had transformed me. I saw the world differently, and I had become a
mathematician. And I was now able to see beauty everywhere around me.

A MIRACLE

But why the Mickey Mouse quotation, I wondered. Gordon explained that his first encounter with Mickey Mouse was in his role as the sorcerer's apprentice in the film *Fantasia*. When his master the sorcerer had gone to bed, leaving his apprentice to carry heavy buckets of water to fill a cauldron, Mickey put on the magician's hat and transformed a broomstick into a legion of water bearers. But when the cauldron was full, Mickey was unable to find the magic formula to tell the broomsticks to stop. A flood ensued, and Mickey was punished. He had transgressed, had been a naughty mouse. But he had dared to attack a difficult problem, and although he got himself into trouble, he went on to conquer the world as one of the best-loved of all creatures, a mouse who brings a smile to every face on the planet at the mere sight of his trademark ears.

Gordon said something like this:

> *Ken, be like naughty Mickey Mouse. You are already a magician creating lovely mathematics. Mickey had an inauspicious beginning as the apprentice. He tried to do something before he was ready. But he ultimately triumphed and became a worldwide symbol of joy and magic. I predict that you, too, will emerge as someone who was meant to follow in your father's footsteps. Your energy and youthful enthusiasm are palpable. I already see it. Now, if you can dream it, you can do it.*

I couldn't believe what I was hearing. His words vanquished the depression that I had brought back with me from Missoula. Right at the moment that the script called for criticism—Ken-chan, you screw up—Gordon offered me the praise that I had so desperately sought as a tiger boy. But it was more than that. He was in effect praising me for screwing up: Ken, you messed up, but that's because you reached beyond your grasp. Now go out and conquer the world. Gordon's words became my battle cry. I had a new voice in my head: "Ken, be like naughty Mickey."

Mathematicians who knew me in the early 1990s will now understand why I attended conferences wearing a stylish baseball cap emblazoned with the image of Mickey Mouse. I didn't wear the cap as a souvenir of Disneyland or because I wanted to be a Mouseketeer. I wore it to remind me of Gordon's inspirational words. That cap was my talisman.

That meeting with Gordon turned out to be one of the most uplifting moments of my life, one whose memory brought me to tears a few years later as I attempted to recount it at a conference I had organized with George Andrews in honor of Gordon's sixty-fifth birthday. I wanted to tell the world how important Gordon was to me.

It was time to get back to work. We had three weeks to prepare for the Rademacher Centenary Conference at Penn State. Some of the most important figures in algebraic number theory would be there: George Andrews (Penn State), Bruce Berndt (University of Illinois), Dorian Goldfeld (Columbia University), Harold Stark (UCSD), among many others. I had learned from the mistakes I had made in Missoula, and together we prepared a talk for this high-powered audience.

We huddled over the coffee table in Gordon's reading room, scribbling an outline, which we then shaped into a well thought out presentation that I later printed on overhead transparencies. When we were done, the room was a sea of paper, as if a powerful box fan had been turned on full blast in front of an unsuspecting stack of recycling paper.

Since Ramanujan scholars Andrews and Berndt would be in the audience, we decided that I should speak about congruences for modular forms, the theory that Deligne, Serre, and Swinnerton-Dyer had developed to explain some of Ramanujan's classic formulas. I had made some small contributions to the field, and using Ramanujan's well-known results as motivation, I hoped to keep the attention of the audience long enough to explain my findings. And if I succeeded, then perhaps I would impress some experts with my results.

I began my presentation with Ramanujan, using his renown as a hook to draw my audience in. Not that I needed a hook. In contrast to my lectures in Missoula, this time I was speaking to an audience of experts in the field, a mathematician's version of "preaching to the choir." My lecture was well received, and I was even invited to submit a paper to the conference proceedings. Several famous number theorists approached me after my lecture, and they congratulated me on the nice work I had done. Although my results didn't astound anyone, I had produced respectable work that experts felt was worthwhile. As the German mathematician Leopold Kronecker wrote in a letter to his colleague Georg Cantor, quoting an old saying, "When kings are building, there is work for carters." He went on to say that a mathematical researcher has to be both king and carter, but at this stage of my career, and especially after the Missoula disaster, I was delighted to earn praise for my yeoman work. After all, even Mickey Mouse had gotten his start as a water carrier. Those number theorists had been kind, and they seemed genuinely to want to encourage me as a junior member of their guild.

In Lutherville after the Rademacher conference (*left to right*: Doug Bowman, Ken Ono, Takasan)

With the success of the Rademacher conference buoying me up, I could look back over the past eight years with a certain satisfaction. I had come a long way, and as I traced my career from high school to college to graduate school, I saw a common thread, and that thread was Ramanujan. When I was a high-school student, Ramanujan, because he had been a hero to my father, had been the Ariadne's thread that allowed me to escape the labyrinth in which I felt hopelessly trapped. Ramanujan's story of achieving success by following his passion even if it meant twice flunking out of college had inspired my father and then me as well. Thanks to Ramanujan, for reasons that I still did not fully understand, my parents had let me run away from my former life. Then when I was drifting as an unmotivated college student, Ramanujan had given me hope, inspiring me, just as my clock was running out, to apply myself to my studies, despite my fear that I would discover that I wasn't good enough to be a mathematician. I did well enough at UChicago to be rescued by Paul Sally, who went out of his way to help me get into a graduate program.

Then as a graduate student, I had followed Ramanujan's lead again. But now I was following Ramanujan's actual mathematics. I presented my work in the inspirational context of Ramanujan's compelling story, and my talk was well received. Each time I looked at a positive event in my life, there was Ramanujan.

I was making progress toward completion of my dissertation, and I had an advisor who was fully committed to me. But that was not enough to subdue the voices in my head. My parents continued to doubt my prospects, and I still yearned for their acceptance and approval. My first published papers and my lecture at the Rademacher conference weren't sufficient to earn their praise. For them, there was only one acceptable outcome—which actually wasn't my goal— a position at the Institute for Advanced Study or at the very least a professorship at a top university.

Even if it had been my goal, the harsh realities of the outside world continued to confirm my feeling of inadequacy. The academic job market was brutal, and I witnessed close friends suffer the indignity of a failed job search. There were a few dozen graduate students at the Rademacher meeting at Penn State, and many of them had presented results that were much more impressive than mine. There would be postdoctoral fellowships for perhaps three or four of them, and tenure-track positions, at schools I had never heard of, for a few others. The odds were not in my favor. There certainly wouldn't be a job for me at Montana, and I was afraid that there might not be a job for me anywhere.

Nevertheless, I had complete faith in Gordon, and I continued to work on my thesis, creating mathematics for its own sake. At least now the negative voices in my head had some competition: "Ken, be like Mickey. If you can dream it, you can do it."

I defended my dissertation in March 1993. I had worked out theorems that I was passionate about. And they were mine. Like a sorcerer, I had conjured up a proof that previously had not existed. And though I was still a sorcerer's apprentice, I didn't have to fear that my theorems would get me in trouble. The only danger was that they might not be good enough. But I had done my best. Following my advisor as a role model, I had become a mathematician who pursued the beauty of mathematics for its own sake, without worrying too much about whether anybody would care about my work. But now it was time to worry. If I didn't want to starve like Ramanujan, I was going to have to get a job. And moreover, without a job, I would be even more at the mercy of my voices. I wasn't confident that my results would pique the interest of many mathematicians, and I feared that I wouldn't get any job offers from research universities. At the time I defended my thesis, I had no job offers, and the annual hiring cycle would soon be over.

My failure to get any offers gave the voices in my head new power:

Ken-chan, your thesis not very good. You say you do math for own sake, but results you get mediocre. You should have been studying all along instead of riding bike and partying. The UChicago junior professor and Montana professor both speaking truth. You wasting everybody's time.

Then ten weeks after I defended my thesis, in June 1993, by which time all the postdoctoral fellowships and tenure-track positions had been filled, I received an unexpected email from Andrew Granville, a British mathematician at the University of Georgia. He was inviting me to apply for a one-year visiting position; the math department at UGA, he wrote, needed one more instructor to cover its fall courses. That email had come out of left field. I hadn't even applied to UGA. But for whatever reason, I was being invited to apply for a job. If I was interested, I should send my CV immediately to his colleague Carl Pomerance, a UGA number theorist well known for his work on prime numbers.

I responded immediately to Granville's request, but I was not optimistic that my application would result in an offer. My two hundred job applications had not generated a single interview, so why would the UGA solicitation yield anything different? Perhaps the same email had been sent to thirty other new PhDs. And suppose I was lucky enough to get the offer. What would be the point of moving across the country—without Erika, who had another year of school— for a one-year job? And what would a one-year stopgap do to improve my future prospects? It seemed much more likely that the job would only delay the inevitable realization that there was no place for me in academia.

Then on the morning of June 23, 1993, my third wedding anniversary, I heard the most incredible news, news that would change my life. Wrapped in only my bath towel, I sat down at my desk in our Santa Monica bungalow to read my email before taking a shower. I was surprised to find dozens of messages. In the early 1990s, before the era of web browsers and the need for spam filters, I received fewer than half a dozen emails a day. Surprisingly, all of the messages had the same subject line.

The story is told that the writers, critics, and actors who gathered daily in the 1920s at the Algonquin Hotel for lunch once held a contest to see who could come up with the most shocking newspaper headline. Dorothy Parker, famous for her acid wit, won with the two words "Pope Elopes." To a mathematician, the subject line of all my emails was more shocking even than that: "Wiles proves FLT."

A few hours earlier, Andrew Wiles, of Princeton University, had announced a proof of Fermat's last theorem at a conference at the Isaac Newton Institute for Mathematical Sciences at the University of Cambridge. His announcement took place in an auditorium a short walk from the same halls that Ramanujan had roamed eighty years earlier.

Fermat's last theorem was surely the most famous open problem in all of mathematics. It all began around 1637, when the French jurist and amateur mathematician Pierre Fermat wrote in the margin of his copy of Diophantus of Alexandria's *Arithmetica* that he had discovered a "truly marvelous proof" of a certain assertion, "which this margin is too small to contain." The assertion is easily stated. It involves only the counting numbers 1, 2, 3, and so on. What Fermat claimed was that for every integer $n > 2$, there are no nonzero integers a, b, and c for which $a^n + b^n = c^n$. Of course, for $n = 2$, there are such numbers a, b, c, such as $3^2 + 4^2 = 5^2$ and $5^2 + 12^2 = 13^2$. Such examples have been known at least since the time of Pythagoras, who lived more than two millennia before Fermat. In fact, there are infinitely many such so-called Pythagorean triples, all given as side lengths of right triangles. What Fermat was claiming was that for every larger value of n, there are no such triples.

That marginal note came to the attention of the mathematical community after Fermat's death in 1665, when his son published a new edition of the *Arithmetica* complete with Fermat's marginal notes. For the next 350 years, mathematicians failed to reproduce Fermat's marvelous proof, and the consensus is that he did not have such a proof. Indeed, although Fermat produced a proof for the special case $n = 4$, the fact that he never mentioned his "marvelous proof" again suggests that he realized that his method for the special case could not be generalized.

Over the centuries, mathematicians chipped away at what came to be known as Fermat's last theorem—so called because it was the last of his claims still lacking a proof. In the process, a great deal of powerful and elegant mathematics was created, comprising whole new branches of the subject. The fame of the Fermat problem grew and grew, until Fermat's last theorem could be found even in the *Guinness Book of World Records*, where it was listed as one of the "most difficult mathematical problems."

Wiles had based his work on Berkeley mathematician Ken Ribet's earlier discovery that Fermat's claim would follow from a verification of the modularity conjecture, a bizarre claim about modular forms and elliptic curves that was first stated by Yutaka Taniyama at the 1955 Tokyo–Nikko conference, which had occasioned the pivotal event in my father's life, being discovered by André Weil. You could say that in the 350 years since Fermat, mathematics had gone a very long distance out of its way to attack the Fermat problem, and Ribet had suggested how it might be possible to come back a short distance correctly. Amazingly, Wiles's proof also made use of Galois representations, the subject I had been studying in my thesis. I couldn't believe it.

The proof of Fermat's last theorem became a major news item. A story appeared on the front page of the *New York Times*, and Wiles was named one of *People* magazine's twenty-five most intriguing people of 1993, alongside the likes of Oprah Winfrey and Bill and Hillary Clinton. I had no idea that these events would soon change my life. After all, I had nothing to do with the proof of Fermat's conjecture.

A few days later, I received an offer from UGA. I suppose that the excitement generated by Wiles's proof of Fermat's last theorem gave a mathematician who studied modular forms, even a lowly one like me, a certain cachet. Emboldened by the Fermat hoopla, I was poised to take the risk and accept the one-year UGA offer. Without any other options, I had a simple choice: accept the UGA offer or leave academia. When I asked Gordon what I should do, he said without hesitation, "Go to Georgia." But I could not think of accepting the offer without a serious discussion with Erika, who would have to stay behind at UCLA to finish her second bachelor's degree. Erika agreed with Gordon, and I accepted the offer that afternoon.

Now in addition to Ramanujan in my corner, I had Fermat. Everyone was suddenly interested in the three topics that made up the title of Wiles's famous talk at Cambridge: "Modular forms, elliptic curves, and Galois representations." Those were all topics that I had studied in my dissertation, and I began to feel that I had expertise that would be of interest to more than just a few of my mathematician colleagues.

Chapter 29

MY HARDY

Athens, Georgia (1993–1994)

I moved to Athens, Georgia, in August 1993. It was only a one-year position, and after that, the future was as uncertain as ever. But I had done it. I had completed my PhD. I was now Dr. Ken Ono, mathematician. To get this far, I had not traveled an easy path, but inspired by Ramanujan and guided by caring mentors such as my brother Santa, Paul Sally, and Basil Gordon, I had achieved an important milestone. I had hoped that my accomplishment would have elicited the recognition and praise from my parents that I had long sought and now felt I merited. After all, earning a doctorate was part of the formula that my parents had laid out for me long ago. It was a major accomplishment, and I had come through. But I received no acknowledgment from them. I didn't bother to go to my graduation ceremony, assuming that they wouldn't want to attend. My father had advised a number of PhD students, and I would have thought that he would appreciate what it meant to end one's apprenticeship and set forth into the world as a certified member of the profession. I suppose he viewed obtaining a doctorate as just one more chore that was expected of you, like brushing your teeth or walking the dog.

I was now twenty-four years old, and I had quite given up hope of ever earning the praise that I had desperately sought earlier in my life. I had developed defense mechanisms to protect my psyche. Basically, I simply buried my former life. My years growing up in Lutherville were consigned to a black hole. I never told anyone about the period of my life before UChicago. I rarely called home, and I rarely communicated with my brothers. Instead, I had Erika nurturing me at home, and I had Gordon nurturing me in my research. At Woodbury University, where I had taught during my last years of graduate school, I was nurtured by Zelda Gilbert, a psychology professor.

© Springer International Publishing Switzerland 2016
K. Ono, A.D. Aczel, *My Search for Ramanujan*, DOI 10.1007/978-3-319-25568-2_29

I had survived by blocking out the past, and I could look forward to the future with at least some optimism. I had not won one of the precious tenure-track jobs or postdoctoral fellowships that would have set me on the path to a research career, but the one-year position that I had been offered was beginning to look like a terrific opportunity. Like Ramanujan, I had been discovered by an English analytic number theorist, my own personal G.H. Hardy. I like to think of the email that I had received from Andrew Granville telling me of the possibility of a job at UGA as something like the letter Ramanujan received from Hardy in response to his "I beg to introduce myself" letter. Excited by Hardy's reply, Ramanujan had proclaimed, "I have found a friend in you, who views my labours sympathetically." Those words echoed in my mind almost exactly eighty years after they were written. And my Hardy—and come to think of it, my Weil as well—would also prove to be my next mentor, nurturer, and friend.

Andrew Granville, a Cambridge-educated number theorist, was a budding star at UGA, and it was he who had arranged the visiting assistant professorship for me. It seems that I had impressed him at a conference earlier in the year, and he wanted to learn about modular forms because of their role in the ongoing work on Fermat's last theorem. Without Andrew's job offer, my career as a mathematician might have been over before it had begun, and now he was offering me something more, the chance to work with him on mathematical research. It was a wonderful opportunity, and I vowed to make the most of it.

With Andrew Granville in 2013

Athens, a charming college town sixty-five miles northeast of Atlanta, was well known for its music scene, as the home of the rock band REM and the hip music venue the "40 Watt" club. Although Erika was unable to join me in Athens for the year, she helped me make the move. We had never driven across the United States, so we decided to take advantage of my move from the West Coast to the Southeast to see a bit of what lies between. We packed the car with a few of my belongings, mounted my fancy Team Miyata bicycle on the roof of our little blue hatchback, and set off.

I wish I could write that the drive was uneventful, but that would be an untruth. The drive was eventful, too eventful. First, we suffered our way across the desert in our little go-kart without air conditioning. Then, somewhere in Arizona, my bike slipped its tether and flew off the roof of the car. I gasped as I watched the reflection of my expensive racing cycle bounce down the highway in the rearview mirror. Thankfully, there were no other cars around. Amazingly, my bike survived its attempted escape.

Near Tupelo, Mississippi, our car began to lose power. It refused to go faster than forty miles an hour. We limped into town and found a service station. The mechanic quickly diagnosed the problem—a malfunctioning catalytic converter. He didn't have the parts; nobody in Tupelo had parts for a Hyundai. The Korean automaker had no presence in the southeast at the time. The mechanic said that he knew a lawnmower repairman who might be able to help us out. A lawn-mower repairman? Was he kidding? Our little car might be a go-kart, but it was no lawnmower. We decided to take our chances on the highway. Poking along at forty miles per hour, we made it as far as Birmingham, Alabama, which became the Hyundai's final resting place. We were 220 miles short of our destination.

I called Andrew from Birmingham and explained that we were having serious car trouble and would likely arrive a day later than we had anticipated. Without a second thought, he offered to make the long drive from Athens to fetch us. I was stunned by his gracious offer, and although we didn't accept it, instead buy-ing an eight-year-old red Honda sports car to complete the journey, I knew then that I would be in good hands at UGA. That road trip was quite an experience, and from it, we learned at least one important lesson: that Honda we bought may have been old, but it had air-conditioning!

Although I was in Athens for only one year, it was an enormously significant time as the last stage of my apprenticeship. Andrew guided my transition from graduate student to young professor, showing what it meant to be a professional mathematician, a colleague who not only does mathematics and teaches classes, but shares in the responsibility of determining the future of the profession and the subject.

Although Andrew was only in his early thirties, he was way ahead of me professionally and served as a perfect role model. He was producing huge volumes of first-rate theorems, covering an unusually broad array of topics: binomial coefficients, combinatorics, graph theory, prime numbers, to name but a few. His work on prime numbers had earned him an invitation to speak at the 1994 International Congress of Mathematicians, one of the most coveted honors awarded to research mathematicians. He was brilliantly advising several PhD students, and I became in effect his postdoctoral advisee. Unlike that unfortunate Hyundai, Andrew was firing on all cylinders. He is the role model that I have done my best to emulate throughout my career.

We spent many hours doing math, gossiping about mathematicians, and enjoying pints of Bass ale at the Globe, the local Bohemian pub popular among the young faculty.

Like Sally and Gordon, Andrew believed in me, and that knowledge was a comfort instead of a source of anxiety. Andrew treated me like a colleague. To my surprise, he gave me early drafts of his papers and asked me to critique his work and his writing. How had I earned the right to question and critique him? He asked me to be a role model to his graduate students. What had I done to merit that? I had been a graduate student myself just a few months earlier. Was I worthy of being addressed as "professor"? Of course not, said the voices in my head; to them, I was still an impostor.

Andrew's unexpected show of respect for my opinions and judgment was eye-opening. It told me that he genuinely believed in my mathematical ability and competence. In a way, that quiet demonstration of respect was the ultimate expression of praise and approval. Although it could never fill the void that only parental praise and approval could satisfy, his confidence in me gave me strength. It was just the boost I needed.

Concerning the right to ask questions, Andrew taught me that it is indeed a right. It isn't a privilege granted only to the select few. It is a practice, he said, that should be second nature to all scientists. Questioning leads to deeper and more meticulous research and to more clearly presented results. It reveals new avenues of inquiry. It is the engine of progress. The ability to ask a question is not something you acquire with seniority. Instead, it is a skill that you must work on at every stage of your development.

That realization gave rise to new voices in my head, voices that asked lots of questions. My self-confidence as a mathematician was born out of those new voices. They told me that I was adequate enough to formulate meaningful questions, and they began to take up the cudgels against the old voices that hammered away at my self-esteem.

Andrew and I wrote one joint paper that year in Athens. It was a semi-important paper in representation theory. We completed a program that aimed to classify the "defect-zero p-blocks for finite simple groups," a project born out of an old problem raised by the Harvard mathematician Richard Brauer in the 1930s. It is interesting to note that Brauer, like Weil, happened to be one of the distinguished American delegates at the 1955 Tokyo–Nikko conference. Our result won me invitations to speak at several first-rate universities.

The most important ingredient in our proof was Deligne's theorem, the work I had studied with Gordon that confirmed a deep conjecture of Ramanujan. I was thrilled to put Ramanujan to good use. His mysterious calculations had inspired the deep mathematics that we needed to crack an unsolved problem in an area of mathematics that didn't even exist during his lifetime. That is the magic of Ramanujan. His formulas are prophecies that have anticipated important discoveries and have guided generations of mathematicians that followed him.

This was the second time that I had been rewarded for following Ramanujan's mathematics. The first had been at the Rademacher conference, where I had presented my results in the context of Ramanujan's formulas from almost a century earlier. Now with Andrew, I had made use of a theorem that solved one of Ramanujan's claims, and we had settled an open conjecture. I was beginning to get the idea that Ramanujan was more than an inspiration in my life. Perhaps I was meant to follow Ramanujan's mathematics, too. He had been the source of everything good in my young career.

Chapter 30

⌒

HITTING MY STRIDE

Urbana and Princeton (1994–1997)

*H*ardy had judged Ramanujan worthy and had invited him into the community of professional mathematicians. Eighty years later, Andrew Granville extended such an invitation to me, and like the earlier pair, we collaborated on research. When we had completed our one joint project, Andrew encouraged me to take aim at some well-known unsolved problems. My year at UGA was an important transition, one that I like to think of as somewhat analogous to Ramanujan's first days in Cambridge. We both had a lot to learn about the world of professional mathematics, and we both had mentors to guide our way.

Andrew helped me spread my wings in search of independence. He told me that I was ready to fly solo and recommended that I study Gauss's class numbers and Euler's partition numbers. Andrew believed that my knowledge and ability could lead me to exciting new results on these two types of numbers. I am pleased to say that I lived up to his faith in me. In a project that would take five years to complete, I obtained important results on Gauss's class numbers in joint work with Winfried Kohnen, of Heidelberg University.

In addition to my research, I was teaching several sections of calculus. I had enjoyed teaching when I was in graduate school, and at UGA, I continued to find teaching rewarding. One thing that struck me right away was that unlike my students in Burbank, who always addressed me as "Ken," here I was "Professor," or occasionally the more informal "Sir." I felt that such formality created a barrier between me and my students, but there was nothing I could do about Southern decorousness, and I must admit that being addressed with such an honorific for the first time in my life reinforced the positive voices that were

© Springer International Publishing Switzerland 2016
K. Ono, A.D. Aczel, *My Search for Ramanujan*, DOI 10.1007/978-3-319-25568-2_30

beginning to do battle against the negative ones. I felt that in teaching, I was giving back something of what I had received from the many wonderful teachers, going back to my earliest years, who had given me so much. I told my students how much it meant to me to be their teacher, and even now, twenty years later, I still hear from some of them from time to time. I have thereby learned that my students take many different paths in life when then leave my class. I heard from one student who is now a physician. Another competed in the 1996 Atlanta Olympics representing Greece. One has become an accomplished botanist. Teaching has become an integral part of who I am.

Now my first semester at UGA was coming to an end, and it was time to think about the next academic year. There had been talk of an extension of my UGA position for a second year, but nothing was certain, and so I was girding myself for another grueling job search. In mid-December, I attended a conference in Asilomar, California, in the hope that my short contributed talk would increase my visibility as I entered the job market. Even if only a dozen or so mathematicians heard my presentation, being noticed by the right person could lead to an offer.

Erika joined me for the conference. We drove up together from Los Angeles, arriving early. Erika understood how important the conference was for my career—I couldn't afford to have a repeat of Missoula—and so she suggested that we take a walk on Asilomar Beach, on the picturesque Monterey Peninsula, a few miles north of the fabled Pebble Beach Golf Course. I needed to clear my mind and mentally prepare myself for the meeting. The cool, stiff ocean breeze refreshed us as we strolled on the boardwalks that crisscross the natural dunes. Erika reminded me that I had done good work, and as a result, good things were bound to happen. Part of me knew that I indeed had begun to make my mark, but hearing it from Erika, I began to believe that I really had something to offer and that I would be able to put that across in my talk. As Erika encouraged me, I heard echoes of Gordon: "Be like Mickey. If you can dream it, you can do it."

Wearing my trademark Mickey Mouse hat, I was standing in line at the conference registration desk waiting to get my information packet when one of the organizers pulled me aside. He told me that the steering committee had decided to offer a presentation that evening, after dinner, on the proof of Fermat's last theorem. Everyone wanted to know about the proof, and in particular, they all wanted to know about modular forms and elliptic curves. Someone had recommended me as the speaker. Would I give a talk on Wiles and Fermat?

Would I? I couldn't believe my good fortune. Although I had had absolutely nothing to do with the proof, I was being given the opportunity to address my fellow mathematicians because my small corner of mathematics had become world news overnight. I took a hot shower, all the while saying, "Oh my God, Oh my God, Oh my God." Then I got to work and wrote the talk. I had learned from my debacle at Missoula to gauge my audience. They didn't want the details of Wiles's proof. They wanted the big picture. I wrote out the main points by hand on overhead transparencies. Everyone was there. My talk was well received, and the next day, my contributed talk, on the rather abstruse topic of Shimura sums related to quadratic imaginary fields, was packed. I began to believe that I might actually have some sort of career as a mathematician.

I devoted the spring of 1994 to the other numbers that Andrew had called to my attention: Euler's partition numbers. Ramanujan was the first mathematician to obtain deep results about them. Although I didn't know it at the time, my decision to work on these numbers marked the beginning of my search for Ramanujan the mathematician. Although I had benefited from following Ramanujan's mathematics twice before, my decision to study partitions was a dive headfirst into some of the deepest waters of Ramanujan's mysteries. Although I had other projects and plans, my commitment to Ramanujan had been set in motion.

The *partition numbers* seem to arise from a child's counting game involving only adding and counting. It is simple to explain what these numbers are about. The equalities $3 = 2 + 1 = 1 + 1 + 1$ illustrate that there are three ways of "partitioning" the number 3. Next, we can observe that $4 = 3 + 1 = 2 + 2 = 2 + 1 + 1 = 1 + 1 + 1 + 1$, which shows the five ways of partitioning the number 4. Repeating this process of adding and counting for an arbitrary number n defines the *partition function* $p(n)$. Thus our two examples are denoted by $p(3) = 3$ and $p(4) = 5$. The partition numbers grow at an astonishing rate. One can calculate $p(10) = 42$, $p(20) = 627$, $p(30) = 5604$, $p(100) = 190,569,292$, and $p(1000) = 24,061,467,864, 032,622,473,692,149,727,991$.

Ramanujan proved very surprising divisibility properties for these numbers. One of his mysterious identities involves the sequence of all partition numbers of the form $p(5n + 4)$. The first few numbers in the sequence are $p(4) = 5$, $p(9) = 30$, $p(14) = 135$, $p(19) = 490$, $p(24) = 1575$. Each of these numbers is a multiple of 5, and what Ramanujan's identity showed is that $p(5n + 4)$ is a multiple of 5 for *every* value of n. He also proved analogous theorems for 7 and 11, namely that for every n, $p(7n + 5)$ is a multiple of 7, and $p(11n + 6)$ is a multiple of 11. These three statements, for the three primes 5, 7, and 11, are now known as *Ramanujan's partition congruences*.

It is natural to ask whether the primes 5, 7, and 11 are somehow special. For the prime 2, for instance, is there a similar progression that yields only even partition numbers? That is, do there exist a whole number A and a whole number

B such that $p(An + B)$ is always even for every value of n? Or is there an analogous progression of partition numbers all of which are divisible by some larger prime number, such as 13 or 677 or 7753?

Ramanujan touched on these questions with enigmatic words in a paper he published in 1919, writing, "It appears that there are no equally simple properties... involving primes other than these three." For decades, mathematicians were unsure what he meant. Did he know about or suspect properties for other primes, but ones that were very difficult to describe? Or was Ramanujan conjecturing that there are no such properties at all for other primes?

Andrew suggested that I study this mystery for the prime 2. In the 1960s, the Canadian-Indian mathematician M.V. Subbarao made Ramanujan's enigmatic claim precise in this case. He conjectured that there is no sequence like those discovered by Ramanujan, that is, no sequence of the form $p(An + B)$, in which all of the partition numbers are even. Although my work fell short of proving Subbarao's conjecture, I proved in the spring of 1994 a theorem that established most of it. The Austrian mathematician Cristian-Silviu Radu would complete my proof fifteen years later. My theorem attracted some attention, and it earned me further invitations to lecture.

The academic year was now drawing to a close. Erika had graduated from UCLA with a bachelor's degree in nursing, and leaving California behind, she joined me in Athens for the summer. It was wonderful to be together again after a year living apart. Back then, the phone company was advertising its long-distance service with the slogan "reach out and touch someone." But long-distance phone calls, as many as we had, couldn't replace our need for each other. I can attest that reaching out and touching are better from up close. I have no idea how Ramanujan managed without Janaki with him England. He must have felt very lonely and isolated.

Erika and I enjoyed Andrew's company and that of the other friends I had made. It was a summer full of hot and muggy bike rides in the hills around Athens, and evenings at the Globe enjoying pints of Bass ale. The summer of 1994 was a very sweet time for us.

Although I had hoped to stay at UGA for the 1994–1995 academic year, budget constraints made that impossible, and I accepted an offer of a visiting assistant professorship at the University of Illinois. It was very difficult leaving Andrew and Georgia after only a year. Through Andrew's wonderful mentoring, I had been transformed into a professional mathematician. But I felt like a bird being kicked out of the nest, and I had doubts whether I could remain aloft and continue to produce results in a new environment without him nearby.

Erika and I moved to Urbana, Illinois, in August 1994. I was delighted to work alongside such fine mathematicians as Nigel Boston, who had arranged the position for me; my friend Doug Bowman; and Bruce Berndt, the Ramanujan scholar whom I had spectacularly failed to impress two years earlier in Missoula. Berndt was devoting his career to resolving all of Ramanujan's claims, and with my newly found passion for Ramanujan's mathematics, I was excited to work with him and his graduate students. In fact, I had gradually formulated a plan to work long-term with Berndt on Ramanujan's mathematics.

Erika and I would spend only one year in Urbana. Erika found a job in the oncology ward of the local hospital, while I spent the year doing mathematics, teaching, and making some difficult academic choices. Although my work on Subbarao's conjecture and my paper with Andrew on Brauer's problem had earned me a measure of notoriety, I had reached a scientific crossroads, a pivotal moment. Shortly after arriving in Urbana, I realized that I had a choice to make.

As college students, future mathematicians are exposed to a wide variety of mathematical subjects. As graduate students, they specialize in some of those subjects in their coursework, and then in their dissertations, they make an original contribution generally in a single specialized area. Early in a career, it is common to pursue questions that are closely related to the work that was done in the dissertation. But at some point, that particular well runs dry or one finds oneself left with questions that are too difficult. Or else one's interest simply turns elsewhere. And of course, some research careers just peter out.

At Urbana, I encountered my own personal crossroads. I could work with Berndt and his students, as I had originally planned, chipping away at the unresolved claims that abound in Ramanujan's notebooks. For if Ramanujan was a mathematical king, he had certainly left behind plenty of work for carters. Indeed, Ramanujan's writings seemed to offer a nearly endless supply of unproven claims and identities to work on. But there was another path I might follow. Following Kronecker's advice, I could choose to take a crack at being my own ruler and assume responsibility for figuring out in what field of mathematics I wished to pitch my tent. Berndt and I had not made any explicit plans for research together, and so I felt no obligation one way or the other. It was like choosing between a promotion to a secure position of responsibility in a large firm or quitting and forming my own startup.

Ramanujan's mathematics, as it is written in his notebooks and letters, presents a major challenge to contemporary mathematicians, who are trained to build frameworks of theory. It is from those new theories that formulas, expressions, and relationships flow. Some mathematicians build theories, while others, the problem solvers, become expert technicians who masterfully apply those theories. There is a similar dichotomy in physics between the theoreticians and the experimentalists. The challenge represented by Ramanujan is that there was no "theory of Ramanujan" to apply to make sense out of his writings. And without such a theory, mathematicians are left to supply proofs to an enormous collection of disparate formulas and claims.

Berndt has chosen to work directly with Ramanujan's writings, seeking meaning in the complicated formulas and expressions one by one. The challenge has been formidable, and together with his students, he had been successfully resolving those beautiful enigmas.

But Ramanujan had become much more to me over the years than the source of enigmatic formulas, and so my view of him and his mathematics began to seek a wider horizon. While studying Ramanujan's claims for their own sake would have been an interesting and important challenge, I had come to suspect that many of those claims could be seen as enticing hints at theories that were just begging to be discovered and developed. I believed that Ramanujan's discoveries had come to him as fragments of a vision of something higher, and so discovering those theories would ultimately give rise to new mathematics. I viewed Ramanujan's claims as gifts for intrepid mathematicians of the future, elusive gifts that concealed their true reward, an offering to any mathematician able and willing to plumb their deeper meaning.

I first began to suspect that there were such unexplored depths in Ramanujan's mathematics when I was working with Gordon at UCLA. Ramanujan's strange formulas had offered hints of the important theories that Deligne had developed. Indeed, Deligne was awarded the Fields Medal for proving overarching conjectures in arithmetic geometry put forth by Serre and Weil. Those theories that arose from Deligne's work have created a framework that has defined a massive part of mathematics for the last forty years. For example, without Deligne's work, we still wouldn't have a proof of Fermat's last theorem. The conjectures Deligne proved were inspired by some of Ramanujan's strange claims for his tau function, claims that seemed insignificant when he first recorded them a century ago.

Instead of working with Berndt, I began, with Nigel Boston's help, to investigate areas of mathematics that although new to me, were known to involve modular forms, the objects I had studied in my dissertation. Instead of following the path of least resistance, I chose to broaden my fields of expertise with the idea that I was investing in my mathematical future. I worked to assemble the knowledge and tools that I believed would be essential for my mathematical purpose: to become one of the heirs to the rich legacy that Ramanujan had left to mathematicians of the future, a legacy available to anyone with the talent, dedication, and insight to seek out the theories implicit in Ramanujan's bequest to the future: the claims he recorded without proof in his letters and notebooks.

I had a strong sense that Ramanujan's claim about partitions in his 1919 paper in which he proved the stunning divisibility patterns for the primes 5, 7, and 11 was such a gift. He wrote, "It appears that there are no equally simple properties ... involving primes other than these three." He didn't elaborate on what he meant. Did Ramanujan know of other properties that were more complex? First of all, if there had been other simple properties, Ramanujan's genius would have found them. And if he had discovered more complex properties, he surely would have written them down. But this was Ramanujan, and there was another possibility: that he sensed the presence of other properties but couldn't see precisely what they were. I read and reread that paper many times, and I almost came to believe that those words were meant for me. In reading between the lines, I became convinced that he had been aware of other, less simple, properties. He was speaking to me, and he was beckoning to me to find them.

I now had new voices in my head, and they were the words Ramanujan had left for me and mathematicians like me. They were the clues he had left behind for us. Ramanujan was telling me that what he had seen in his visions was a fragment of something larger. My mathematical search for Ramanujan now became a search for an encompassing theory.

That year in Urbana, I wrote further papers in arithmetic geometry and representation theory, subjects that at first glance have nothing to do with Ramanujan's mathematics. I was beginning to develop my view of the implications of Ramanujan's mathematics, which I would later write about in a book I called *The Web of Modularity*. The functions I had studied in my dissertation, the so-called *modular forms*, seemed to appear in so many different areas, forming a web of interconnected subjects, that they must have a deep mathematical significance. Thus it was that during my year in Urbana, I developed a personal relationship to Ramanujan's mathematics, and the work I did in recognizing the many implications and roles for those functions, constructing the web of modularity, has driven and sustained my career—a mathematical search for Ramanujan.

In November, out of the blue, I received a letter from the Institute for Advanced Study in Princeton. Nothing about the envelope suggested that the letter inside would be important. It was an ordinary white office envelope with Institute for Advanced Study as the return address. It could have been an announcement of the seminar schedule for all I knew. But it was much more than that. I was like the boy Charlie from Roald Dahl's children's book *Charlie and the Chocolate Factory* when he tore the wrapper off his Wonka bar to discover the last of five golden tickets, offering a tour of the magical factory and a lifetime supply of chocolate. Instead of all the chocolate I could ever want, I had been given an even sweeter prize. The letter was from the renowned number theorist and Fields medalist Enrico Bombieri. It consisted of a single paragraph offering me a two-year membership at the Institute.

Andrew Granville had recommended me to Bombieri, praising my goal of searching for Ramanujan's number theory, and Bombieri had apparently found my goal worthy of support. I had only one duty—to pursue my search for Ramanujan. The letter was a dream come true. I would have the privilege of working at an institution made famous by the likes of Einstein, Dyson, Gödel, Oppenheimer, and Weil, among many other luminaries.

And of course, the Institute had played a special role in my family history. André Weil, who had discovered my father in Japan forty years earlier, was a longtime faculty member. Over the years, he had arranged several visiting positions for my father, including the 1968–1969 academic year, the year I was born. My mother would take long walks on the Institute campus, pushing me in a carriage. Erika and I would have our first child, Aspen, at the Institute in 1996, and history would almost repeat itself: instead of long walks pushing a carriage, Erika and I would glide around the grounds on roller blades with Aspen strapped safely in a pink baby jogger.

Thanks to the strong support of my mentors Sally, Gordon, and Granville, I had somehow reached a level beyond my wildest dreams. I had dropped out of high school ten years earlier, and four years later, my complex analysis professor at UChicago had tried to talk me out of pursuing a doctorate. Now I was pursuing my own research program at the Institute for Advanced Study. And woven through all that history was my faithful guide Ramanujan.

My decision to search for Ramanujan the mathematician would mean going a bit more distance out of my way. I wanted to increase my knowledge, and that would slow down my publication rate, something that anyone trying to land a permanent job must take into account. The Institute's offer was therefore a godsend. I could concentrate on my mathematics in an environment free of other distractions and responsibilities; I had access to world-class libraries at the Institute and Princeton University, and I would be able to learn from some of the world's most brilliant and talented mathematicians. I vowed to make the most of this special opportunity.

Erika and I moved to Princeton in August 1995, my third consecutive August move. We lived in a two-bedroom Bauhaus apartment at 69 Einstein Drive. Erika found work as a nurse in Trenton, and I did my number theory research.

My first task was to complete my web of modularity. After that initial investment, I would then turn to my search for Ramanujan's mathematics.

The 1995–1996 academic year at the Institute was devoted to an examination of the proof of Fermat's last theorem. Although it turned out that the original proof by Wiles had a flaw, he was able to correct it in a supplementary paper written with his former graduate student Richard Taylor. Due to the importance of their work, the Institute had invited many of the world's leading number theorists to spend the year in a collaborative environment in which they could push number theory even further. It was an awesome year, one that would contribute to my growth as a mathematician and help my career in many ways. With so many experts to talk to, I was able to complete the bulk of the work for my web of modularity in good order.

We made many friends that year, mostly other young mathematicians. Two of our closest friends were Princeton graduate students Kannan "Sound" Soundararajan and Chris Skinner. Sound would later become a professor at Stanford, and Chris would become a professor at Princeton. We were fans of the TV show *X-files*, and we made frequent trips to nearby Iselin, New Jersey, for Indian food. Chowpatty was our favorite restaurant, and that is where I developed a taste for south Indian vegetarian dishes like pav bhaji and masala dosa.

My enthusiasm for cycling rubbed off on Chris and Sound. I helped them shop for mountain bikes, and we rode often on the trails in the area, even after one of us flipped over the handlebars on a narrow and rocky descent and landed in the emergency room. When our daughter Aspen was born in June 1996, Chris and Sound became her first "uncles."

I wrote papers with Chris and Sound during my Princeton years. Chris and I would ultimately write three papers on mathematics related to the Birch–Swinnerton-Dyer conjecture, one of the notorious "Millennium Problems" whose solution would bring a million-dollar prize. This was part of filling in my web, the groundwork I felt was necessary before I could begin my search for Ramanujan's mathematics in earnest.

When I wasn't thinking about that problem, Ramanujan's words were on my mind, as if he were somehow speaking to me. In addition to his comments about the absence of further "simple properties" for the partition numbers, I was deeply interested in a 1916 paper on quadratic forms in which similar puzzling words appeared, this time about the absence of a "simple law." Sound and I became enamored with the problem implied by Ramanujan's words. We had to figure out what he meant.

In some ways, Sound is a modern-day Ramanujan. Born to Brahmin parents in Chennai (Madras), he discovered mathematics as a young boy. He was a prodigy, and he came to the West seeking to make a name for himself. As a high-school student in 1989, when I was racing my bike against people like Greg Lemond and Eric Heiden for Pepsi-Miyata, he attended the Research Science Institute at MIT, arguably the most renowned summer science research program for high-school students. There he began to hone his skills in analytic number theory. In 1991, he won a silver medal representing India at the International Mathematical Olympiad in Sweden. He attended college at the University of Michigan, where he wrote an honors thesis that earned him the prestigious Morgan Prize for undergraduate research in mathematics.

Sound and I wanted to figure out what Ramanujan had meant by the absence of a "simple law" for his quadratic form. Quadratic forms are objects that mathematicians have studied for centuries. One of the most famous theorems about them is due to the eighteenth-century mathematician Joseph Lagrange. He proved that every positive integer—no exceptions!—can be expressed as the sum of four perfect squares. It's like a magic trick: Pick a number, any number. How about 374? Then I can pull four integers out of my hat such that their squares add up to 374. For example, $374 = 0^2 + 2^2 + 9^2 + 17^2$. I could also have written $374 = 6^2 + 7^2 + 8^2 + 15^2$ (as you can see, there is nothing necessarily unique about such representations). Lagrange proved that there is nothing special about 374. You can find a similar representation for *every* positive integer.

Now in solving a mathematical problem, it is crucial to ask the question in a way that leads to a solution. So instead of asking whether every positive integer can be written as a sum of four squares, Lagrange considered the *quadratic form* $a^2 + b^2 + c^2 + d^2$ and proved that by plugging in all possible integer combinations for a, b, c, d into that quadratic form, you obtain all of the numbers 1, 2, 3, 4, ...

In the 1916 paper that had intrigued Sound and me, Ramanujan was considering the quadratic form $x^2 + y^2 + 10z^2$, in relation to which he wrote, "the odd numbers that are not of the form $x^2 + y^2 + 10z^2$, viz., 3, 7, 23, 31, 33, 43, 67, 79, 87, 133, 217, 219, 223, 253, 307, 391 ... do not seem to obey a simple law." Whereas Lagrange's quadratic form could represent every positive integer, there are numbers that cannot be written in the form $x^2 + y^2 + 10z^2$ (the reader can easily verify that the numbers in Ramanujan's list above cannot be thus represented).

There are plenty of odd numbers that *can* be obtained by this quadratic form, such as $57 = 1^2 + 4^2 + 10 \times 2^2$, where we have chosen $x=1$, $y=4$, and $z=2$. There seemed to Ramanujan to be no simple law that would explain his list and show how it continues. What did he mean that there doesn't appear to be a "simple law"?

In 1990, Bill Duke and Rainer Schulze-Pillot proved a fantastic theorem that implied that all odd numbers *from some point on* must be represented in this way. That meant that Ramanujan's list petered out eventually, since the number of odd integers that cannot be represented by the quadratic form in question is finite. You could say, then, that Ramanujan was right: there was no simple law. Indeed, you could say that there was no law at all! It was just a finite list of the relatively few (however many it might be, a finite number is small compared to infinity) integers that happened not to have such a representation. On the other hand, for us, the law that we had to find was this: what is the last number on the list? And was finding it going to be simple, or was it going to be hard?

Sound and I ran a computer program, and we found that Ramanujan's list could be extended by the odd numbers 679 and 2719. But after that, the well ran dry. For every larger odd number we tried—and we tried all the way up to the super-huge number 1,000,000,000,000,000—we found that it *could* be expressed by the quadratic form using some choice of x, y, and z. We concluded that we had found the "from some point on" from Duke and Schulze-Pillot's theorem. That was the "simple law." It must be true that every odd number larger than 2719 can be expressed by Ramanujan's quadratic form, and we set out to prove it. Although we firmly believed that we were right, we couldn't come up with a proof, no matter how hard we tried. When mathematicians are unable to prove something that they believe to be true, they sometimes are able to give a proof on the assumption that some unproven conjecture is true. In our case, we were able to prove that 2719 is the last number in the sequence on the assumption of the truth of the *generalized Riemann hypothesis.*

The ordinary Riemann hypothesis is one of the most important open problems in mathematics. It involves a certain conjectured property of a certain function of a complex variable. Its truth would resolve many unanswered questions. For example, mathematicians would have a much clearer understanding of prime numbers if the Riemann hypothesis were confirmed to be true. The generalized Riemann hypothesis is a natural generalization of the Riemann hypothesis.

Using a long and complicated argument, we finally found a way to show that the truth of the generalized Riemann hypothesis implies that every odd number greater than 2719 can be written as $x^2 + y^2 + 10z^2$ for some integers x, y, and z. The fact that almost every mathematician believes in the truth of the generalized Riemann hypothesis and the fact that every odd number greater than 2719 up to a very large number can be represented by Ramanujan's quadratic form convinced us that we had found the law. But although the law is simple enough to state, it thus far defies a definitive proof. To be sure, if someone manages to prove the generalized Riemann hypothesis, then our conditional proof will at once become a genuine proof. But the generalized Riemann hypothesis is arguably one of the most difficult open problems in mathematics. So Ramanujan was right that the odd numbers do not obey a simple law, in the sense that they are constrained by one of the most difficult unsolved problems in mathematics.

I had no idea that I would see the number 2719 again ten years later, etched on a wall in the very spot where Ramanujan performed some of his first calculations.

Thanks to my two years in Princeton, I was able to complete a large body of work, and I was rewarded for my efforts with a tenure-track assistant professorship at Penn State University. George Andrews, one of the mathematicians I admired from the BBC documentary and Robert Kanigel's book *The Man Who Knew Infinity*, had recommended me for the position. My work on Ramanujan's mathematics, namely my work on Subbarao's conjecture for the partition numbers and my work with Sound on Ramanujan's quadratic form, were the reasons for my success.

Chapter 31

⌒

BITTERSWEET REUNION

Princeton, New Jersey (1997)

S hortly before we moved to Penn State, my parents made the drive from
Baltimore, where my father was still teaching at Johns Hopkins, for a
weekend visit. I had arranged for them to be reunited with André Weil,
who was now ninety-one. My parents had not seen Weil since the 1970s, and
they didn't know what to expect. They were nervous. However, the reunion was
important to them. They needed to thank him for his generosity, for having
made their lives possible.

Watching my parents walk the grounds of the Institute and hearing them
retell the story of how André Weil rescued them as a young starving Japanese
couple in 1955 was profoundly moving. Weil had invited my father to the
Institute, and the hallowed grounds now represented the beginning of everything that was good in their lives. They excitedly pointed out where they used to
go for walks, and they jokingly pointed out the "Tamagawa tree," a tree on
Einstein Drive that Tsuneo Tamagawa, their friend and future Yale professor,
had rammed with his car while learning to parallel park.

Sitting in the Institute's Fuld Hall waiting for Weil to arrive, my father talked
about the Tokyo–Nikko conference. I knew only a small part of the story, the bits
that most professional number theorists know. But there was more that my
father wanted to tell me.

At that conference, Weil gave an impromptu after-dinner talk that wasn't part
of the official program. He wanted to inspire the young Japanese mathematicians
by telling a story that had inspired him when he had been uncertain of his future
in mathematics.

© Springer International Publishing Switzerland 2016
K. Ono, A.D. Aczel, *My Search for Ramanujan*, DOI 10.1007/978-3-319-25568-2_31

Weil had a fascinating life, in a way reflecting that of my father. Although he was born into a prominent Alsatian Jewish family and grew up in privilege in Paris, he had struggled early in his career.

As a young mathematician, Weil traveled to India and soaked up the culture—not many Western mathematicians were doing that in the early twentieth century—and throughout his life, he was influenced by Hindu thought. Weil was traveling in Scandinavia in 1939 and was in Finland when war broke out between Finland and Russia. He was mistakenly arrested by the Finns for spying. He was soon released, but on his return to France, he was arrested for failing to report for military duty, and he spent time in prison. In 1941, he managed to sail with his family from Marseille to New York, and for several years he was a mathematical nomad, struggling in his own kind of purgatory for a mathematician of his stature.

Weil was saddled with a heavy teaching load at Lehigh University for two years, followed by two years in a low-paying position at the Universidade de São Paulo, in Brazil. His troubles were somewhat reminiscent of Ramanujan's own struggles. Finally, the importance of his work was fully recognized, and he was suitably rewarded with a faculty position at UChicago in 1947.

Inspired by the purpose of the symposium—the promotion of reconciliation and world peace in the name of science—Weil felt the need to tell a story that had inspired him when he needed hope. Dressed in a white shirt and light khaki pants, Weil addressed the Japanese mathematicians, almost all in dress shirts, neckties, and dark suits, politely sitting in formation at tables behind cardboard rectangles bearing their names.

He spoke about a most enigmatic man, an untrained college dropout from faraway India, who had overcome unbelievable obstacles forty years earlier to startle a few of the world's leading mathematicians with his creativity and imagination before his untimely death at the age of thirty-two. The young Japanese men had never heard of that man, Srinivasa Ramanujan, but by the end of the evening, they had all embraced a new hero. Ramanujan had inspired Weil, and that night, he inspired those young Japanese mathematicians, hungry in body and spirit, who needed hope and inspiration.

André Weil lecturing at the 1955 Tokyo–Nikko conference

Oh my God, it all made sense now!

It was no wonder that my father had been so overcome by Janaki's letter, though it was little more than a form letter that she had sent to dozens of mathematicians. I now understood that on that fateful day in 1984, Janaki's letter had revived within my father memories of his own life's pivotal moment. Ramanujan had been an inspiration to him, a mathematical hero who embodied everything admirable in an enquiring mind pitted against hardship. And he had learned about Ramanujan at the conference where he himself had been discovered by the great André Weil, setting in motion everything that had been good in his life. Janaki's letter had carried my father to that moment in 1955 when it was revealed that all his efforts had not been in vain, that he, too, counted.

The clouds parted; the sky was suffused with light; somewhere, an angelic choir was singing; the voices in my head fell silent. I saw before me my destiny and purpose. I was to be the next link in the chain. I was to lead a life that mirrored my father's path. We had both earned positions at the Institute after overcoming formidable challenges. And it turned out that we had both benefited from strong mentors who helped pave our way. Had it not been for Weil's generosity, what would have become of my parents? Had it not been for the strong support of men like Sally, Gordon, and Granville, and had it not been for Ramanujan and the "Ramanujan miracles" that seemed to occur when I needed them most, what would have become of me?

When I offered the example of Ramanujan in a last-ditch effort to convince my parents to let me drop out of high school, I did so because I was grasping at straws. But in uttering the Open Sesame of Ramanujan's name, I unwittingly did much more than secure my freedom. I offered my father a glimpse of the next chapter in an infinite story that continues to evoke awe and wonder.

Janaki's letter had reminded him that not all paths can be safely laid out in advance. And that recollection allowed him to see me in a different light. Perhaps I could find my own way toward living the life that was meant for me, a life that made good use of my talents. It was that glimpse of possibility that convinced him to let me go. The enigmatic Ramanujan had helped make my father, and because of that, he was somehow helping to make something of me.

The end to the weekend was bittersweet. The ninety-one-year-old Weil finally arrived. Although he had long since retired, he still spent much of his time at the Institute, sitting alone in a comfortable chair in the tea room of Fuld Hall. He rarely spoke to Institute members. He kept to himself, and he seemed perfectly content in doing so.

Weil was now nearly deaf and blind. He was unable to focus his gaze in the direction of my parents as they addressed him. It was clear to me that Weil remembered my parents, and I am certain he understood something of what they were saying. However, all he could muster in return, in a soft melancholy voice, were the words, "I am so sorry Ono. I cannot hear nor see you." And that is how it ended.

In tears at the sight of this once powerful man, they left Weil alone in his comfortable chair, staring off into space. The man who had been instrumental in virtually everything important in our lives, who had been a towering figure in twentieth-century mathematics, had been reduced by age and infirmity to a mere shadow of his former self. Weil would pass away within the year.

That weekend was the first time that I was able to offer my parents convincing evidence that I had achieved a level of success of nonzero measure. I had become an adult in their eyes and was no longer disparaged as a high-school dropout and unmotivated college student. It didn't matter that I had quit the violin, nor that my complex-analysis professor felt that I had insufficient talent. It was of no importance that I had almost failed my algebra qualifying exam. That my first two hundred job applications had resulted in zero interviews was a thing of the past. Those events, which had been delicious nourishment for the negative voices in my head, had lost their hold. I was hosting my parents as a member of the Institute for Advanced Study, Einstein's institute in the woods. I had reconnected them with Weil, allowing them to recall all the contingencies of their own journey from war-torn Japan to the Baltimore suburbs.

That weekend marked a turning point in my relationship with my parents. They now understood that I was developing a strong reputation as a mathematician. They also enjoyed their role as grandparents to Aspen, even pushing her around the Institute grounds in a stroller, reliving the days thirty years earlier when I was the one being pushed. As if a wicked sorcerer's spell had been lifted, my parents stopped worrying about me. It was a huge relief to me to realize that I had achieved a measure of professional success in their eyes. They stopped voicing their disapproval of me, no longer harping on my perceived inadequacies and failures.

Since that moment, they haven't missed holidays, birthdays, and anniversaries—all the celebrations that were absent in my previous life. I can now count on my mother to send cards for each of those occasions, just as I could count on her setting out my breakfast all those years ago. Years later, when they are seventy-eight and eighty-six, my parents will even travel to Atlanta to celebrate Aspen's high-school graduation.

Chapter 32

~

I COUNT NOW

State College and Washington, D.C. (1997–2000)

*A*spen, Erika, and I moved to State College, Pennsylvania, home of Penn State, in July 1997. To celebrate my appointment, George Andrews and I organized the conference "Topics in Number Theory in Honor of Basil Gordon and Sarvadaman Chowla." The meeting was held from July 31 to August 3 at the Penn State Hotel. George and I wanted to honor the memory of Chowla, a longtime member of the Penn State faculty who had passed away in 1995, and we wanted to celebrate Gordon's sixty-fifth birthday.

It was a glorious meeting. Over 170 number theorists from all over the world attended the conference and joined our celebration. We heard about cutting-edge research from such stars as Henri Darmon (Princeton), Richard Stanley (MIT), and Trevor Wooley (University of Michigan), in addition to my friends Andrew Granville, Carl Pomerance, Chris Skinner, and Kannan (Sound) Soundararajan.

I couldn't think of a better way to celebrate Gordon's birthday than to offer him this conference in his honor. Gordon had taught me how to love math for math's sake. He had offered me strength when I was at my lowest point.

I had prepared a moving speech for the conference banquet in which I would thank Gordon for everything he had done for me—showing me how to see beauty in mathematics, helping me to overcome my insecurities, and believing in me when I didn't believe in myself. I wanted him to understand how important he had been in my life. I spent days working on that speech, discarding perhaps a dozen drafts. I agonized over the words. I had to get it right. Finally, I pared away all the high-sounding periphrasis and circumlocution and wrote a simple three-minute talk straight from my heart.

© Springer International Publishing Switzerland 2016
K. Ono, A.D. Aczel, *My Search for Ramanujan*, DOI 10.1007/978-3-319-25568-2_32

It was the night of the banquet. When it was my turn to speak, I walked to the podium and unfolded the paper on which my speech was written. Standing before family, friends, and 170 fellow number theorists, I broke down and cried. I tried to find the strength to read my speech, but I was so overwhelmed by the emotions stirred up at the sight of its first words that I couldn't even begin. When it became apparent that I was not going to regain my self-possession, that I was going to continue to stand there, choked up, unable to utter a word, Gordon walked up to the podium and embraced me. He didn't need to hear my words. He knew, and everyone knew. I might have saved myself the trouble of writing a speech and simply borrowed that overused phrase, "Words cannot describe what Basil Gordon means to me." I am crying as I write this.

I handed the folded paper to Gordon. It began with the words, "I thank Basil Gordon for saving my life."

Basil Gordon passed away in 2012 after a long, happy, and fulfilled life. I was honored that his family asked me to speak at his memorial service. This time, in front of his family and former UCLA colleagues, I fought through tears, tears shared by everyone there, and I finally delivered my speech. I miss him deeply, and I thank God for sending me Basil Gordon.

Shortly after the conference, I received an exciting offer from Bruce Berndt. Aware of my plan to search for Ramanujan the mathematician, with the idea of seeking the theories of which his claims offered enticing glimpses, Berndt asked me to help him edit one of Ramanujan's unpublished manuscripts. The manuscript in question was an extensive collection of notes on Ramanujan's tau-function and Euler's partition numbers. We planned to publish a paper under Ramanujan's name, to which we would add commentary for interested readers.

I had already planned to study Ramanujan's writings, looking for the clues that I felt had been placed there for me to discover. The proposed project with Berndt offered the perfect opportunity, and it involved one of Ramanujan's unpublished manuscripts.

Most of these unpublished notes involved ad hoc calculations with modular forms, and so Bruce thought of me as a natural candidate to help him with the task. I could provide a modern perspective, one based on the ideas of Deligne, Serre, and Swinnerton-Dyer. I had no idea that Bruce's request would lead to insights that would give an enormous boost to my career.

It was exciting to read Ramanujan's unpublished notes. His 1919 observation, "It appears that there are no equally simple properties ... involving primes other than these three," had been haunting me for two years. The "simple properties" to which Ramanujan referred were his partition number patterns for the primes 5, 7, and 11. Did he know of properties for other primes? If so, were the patterns complicated? Apart from my work on Subbarao's conjecture, which concerned even and odd partition numbers, and work of A.O.L. Atkin from the 1960s, virtually nothing was known about these questions. It was from reading Ramanujan's unpublished manuscript that I began finally to see what Ramanujan probably had in mind.

My Penn State office was on the top floor of the McAllister Building, a former women's dormitory that was apparently no longer fit to house undergraduates but was considered good enough for mathematicians. Built in 1904, the building was in desperate shape. I could hear squirrels rummaging in the attic above my office. The Internet went out several times during my first year: it seems that squirrels like the taste, or at least the mouthfeel, of cables. I had a window whose lock had been torn off years earlier. Fortunately, there was no way of opening it, since it was permanently sealed shut with layer upon layer of heavy white paint.

I furnished my office at Penn State Salvage, a warehouse of castaway odds and ends scavenged from the dorms. I bought a stained freakish orange sofa for $20, and I placed it under the low ceiling that followed the gabled roofline of the building. It was on that sofa that I did much of my best work at Penn State, and it was there that I came across a few formulas related to partition numbers in Ramanujan's unpublished manuscript that made no sense. I had been thinking about partition numbers for three years, and I became convinced in looking at his formulas that Ramanujan must have made some errors in his calculations. But I was unable to pinpoint them. I then decided that I was probably misunderstanding what he meant, which is common for anyone reading Ramanujan's writings. I didn't even see how to compute his expressions. His formulas simply looked too strange to be true.

I had been pondering those bewildering expressions for an hour when a glorious insight came to me as if handed down from on high. It was one of those inexplicable moments of clarity that come in a flash when one is in a deep meditative state. It was one of those moments when the clear outline of a solution emerges from the fog.

I excitedly jumped to my feet, banging my forehead so hard on the sloping ceiling that I sent the squirrels in the attic scurrying. I was stunned by the blow, but my mind was clear: I understood what Ramanujan meant by his expressions, and I knew how to compute them. Ramanujan had found a way to relate the partition numbers to special functions I already knew well. It was time to do some calculations. I went to my computer and wrote a little program to evaluate one of the formulas. The computer spit out the terms one by one, and the results showed that the partition numbers were in perfect harmony with the values of the special function. Ramanujan was right, and I had figured out his secret.

I ran out of my office wanting to hoot and holler in sheer joy. But I thought better of that and ran to the bathroom to splash some cold water on my face to make sure I wasn't dreaming. I was so excited that I drenched my shirt, which I then took off. At that moment, in walked Dale Brownawell, a fellow number theorist, to find me shirtless, wet, and sporting a huge bump on my forehead. Understandably shocked by the sight of me, he said, "What in the world happened to you?" I didn't want to have to explain my "eureka" moment on the sofa, since there was still work to be done before I would have anything to say to anyone, and so I simply said that I had hit my head on the coat hook in the bathroom stall. I was not very convincing and probably gave Dale cause for unnecessary speculation.

I spent the next weekend pondering Ramanujan's strange formulas, now with the firm belief that they had to be correct, "because if they were not true, no one would have had the imagination to invent them," as Hardy said after receiving his first letter from Ramanujan. After trying a variety of tricks from the theory of modular forms, I was finally able to devise a procedure that produced Ramanujan's formulas and thereby proved his claims. From those puzzling claims, which Ramanujan had offered without proof in his strange unpublished manuscript, I assembled a theory, which I was immediately able to put to good use.

By combining my new results with work of Deligne and Serre and results of Shimura, one of the stars among the young Japanese mathematicians at the 1955 Tokyo–Nikko conference, I was finally able to make sense of Ramanujan's enigmatic quotation. I proved that the partition numbers have patterns for virtually every prime. For primes larger than 11, however, those patterns turn out to be monstrosities. For instance, the number of ways of partitioning $4,063,467,631n + 30,064,597$, no matter what you choose for n, is always a multiple of 31. I proved that there are infinitely many such patterns for every prime other than 2 and 3.

Ramanujan was right about there being no simple properties for primes larger than 11, and I am certain that his unpublished formulas were the first steps in his effort to prove the complex properties that he knew must be out there. Moreover, I firmly believe that he expected that he would eventually find the necessary tools for proving them. He never succeeded. That I was able to succeed where he had failed is because I had at my disposal the powerful mathematical machinery assembled fifty years after Ramanujan's death by Deligne, Serre, and Shimura. When I proved my theorem, I was standing on the shoulders of giants.

My paper proving this theorem appeared in the *Annals of Mathematics*, the same journal that had published the proof of Fermat's last theorem just a few years earlier. My result made world news, and I was rewarded for figuring out Ramanujan's enigma with fellowships from the Alfred P. Sloan Foundation and the David and Lucile Packard Foundation. I was also one of sixty young scientists and engineers to be honored by President Bill Clinton with a Presidential Early Career Award. My parents attended the ceremony at the White House, and I was proud to have them there.

That evening, after the White House ceremony, my father bestowed on me the magical letter that Ramanujan's widow had sent him in 1984, the one that had come to symbolize my path from high-school dropout to professional mathematician. He said,

> *Ken-chan. Ramanujan's widow sent it to me for the little part I played in honoring his memory. Your work, that is the genuine gift. There is no better way to honor Ramanujan, and my son, you were the one to do it. I was merely meant to be a temporary keeper of the letter, and now I pass it on to you, its rightful owner. I am so proud of you.*

I finally heard the words I had so desperately craved my entire life. That evening, I cried tears of joy under a steaming hot shower, realizing that I had finally achieved my impossible dream. At that moment, many of the voices in my head vanished. I have never heard from them again. Although I still have some voices, they are of the sort that everyone has. The voices that had once nearly driven me mad disappeared that day a few blocks from the White House. As fate would have it, I was thirty-two years old, the age at which Ramanujan, the "gift from Kumbakonam," passed away, leaving behind writings that have been speaking to me.

In front of the White House
before my Presidential Early
Career Award ceremony

Chapter 33

~

THE IDEA OF RAMANUJAN

*H*ow have I been lucky enough to get to where I am today?
The story of Ramanujan has inspired generations of mathematicians. It inspired Weil. It inspired my father as an uncertain mathematician in postwar Japan, and in turn it inspired me as a troubled teenager in the 1980s.

My search for Ramanujan shall go on. I shall continue to search for Ramanujan the mathematician. Perhaps more importantly, I shall continue to promote the "idea of Ramanujan"—that greatness is often found in unusual and unpromising circumstances, and it must be recognized and nurtured. Indeed, had it not been for the goodwill of Ramanujan's friends and parents, and people like G.H. Hardy, André Weil, Paul Sally, Basil Gordon, Andrew Granville, Bruce Berndt, George Andrews, and countless others, I wouldn't have had anything to write about. Ramanujan, my father, and I—we three would have never happened.

I will continue mentoring bright math students, the future Ramanujans, through my annual summer undergraduate research programs, my training of PhD students, and my mentoring of postdoctoral students. What is interesting about many of the students that I have mentored, like the prodigies Jayce Getz, Daniel Kane, Eric Larson, Hannah Larson, Alison Miller, Maria Monks, Evan O'Dorney, Aaron Pixton, among others too many to mention, is that they have often come from nontraditional or unpromising circumstances. What unites them is that they have been drawn by the beauty of mathematics. I owe it to them and my mentors, and I owe it to Ramanujan, to do my part. Searching for Ramanujan is my calling; it is my life's purpose.

How was I lucky enough to get to where I am today? That is the question that opened this book. The preceding pages prove that the answer is multidimensional, and the many steps in the proof amazingly trace my search for Ramanujan. I benefited from the tough love of my parents. They fostered qualities in me that

© Springer International Publishing Switzerland 2016
K. Ono, A.D. Aczel, *My Search for Ramanujan*, DOI 10.1007/978-3-319-25568-2_33

have been essential to my success. I am ambitious and competitive. I am restless, anxious to take on the next challenge.

But I also emerged as a fragile adolescent with low self-esteem. I needed nurturing from many people along the way to reach my happy place, and for some inexplicable reason, Ramanujan was there with me every step of the way.

I recently gave a TEDx talk that I called "How to live mathematically, but not by the numbers." As the product of tough-loving Japanese-American parents, and as a professor at Emory who has present-day tiger children in class, I felt the strong need to offer advice.

Let me offer this advice here.

How will you choose to live your life?

That is the question I posed in my TEDx talk. It is the most important math problem that most people are ever asked to solve. It is a question that researchers have studied, and some have even offered mathematical formulas as solutions.

To keep it simple, I maintain that many people today are living mathematically: they "live by the numbers." Many high-school students, college students, and their parents will know exactly what I mean. Kids today are frantically living their lives in pursuit of numbers such as high grade-point averages, strong ACT and SAT test scores, a large number of "Advanced Placement" courses and tests, admission to a top-ranked college, and so on. Many parents, and society as a whole, emphasize the importance of these numbers with the idea that strong numbers lead to success, and success leads to happy and fulfilled lives.

Yet such a frantic high-stakes pursuit has its risks.

But despite those risks, does the frantic pursuit of credentials work? How do we measure happiness and fulfillment? It is natural to ask those who are facing death. Well-known studies have determined the top deathbed regrets, and their findings are thought-provoking. These are the top five deathbed regrets from Bronnie Ware's book *The Top Five Regrets of the Dying*:

1. I wish I had had the courage to live a life that was true to myself, not the life others expected of me.
2. I wish I hadn't worked so hard.
3. I wish I had had the courage to express my feelings.
4. I wish I had stayed in touch with my friends.
5. I wish I had let myself be happier.

These regrets speak to me. It turns out that many of the choices I made in my life have warded off the possibility of such regrets.

I struggled to live my own life. I found strength in my decision to quit the violin, and I found freedom in my bicycle racing. I strongly believe that both of those choices helped me find the courage to live a life that was true to myself. My choices were made in the pursuit of happiness, and it was my ability to express my feelings with conviction that allowed me to make those choices.

I was a goofball in college. I was an unmotivated student who was worried about living up to parental expectations as a math major at UChicago. Although I ended up becoming a mathematician, I did not work on math full throttle, living the life that my parents expected of me. As a result, I did not come to resent mathematics, and I was able to remain open to the idea of pursuing mathematics for its own sake when Gordon taught me how to appreciate its intrinsic beauty. Of course, thanks to that choice, I ended up enjoying many rich collegiate experiences that I would otherwise have missed out on. They were choices made in the pursuit of happiness.

Among the top five deathbed regrets, in case anyone is keeping score, the only one I haven't checked off is number 4: I wish I had stayed in touch with my friends. Writing this book has been a cathartic experience. I have fondly recalled some joyous times while struggling to find the words and courage to reveal my darkest moments. Many of these darkest moments were new to Erika, who has been closer to me than anyone. It is my hope that with this book, I will be able to reconnect with some of my friends from the black hole of my former life.

Friends, I apologize for abandoning you. It is my sincere hope that you will understand from this book why I had to leave you behind for a time. I went a very long distance out of my way to block out the period of my life in which you were present, and I am ready now to come back the short distance that separates us.

Let me end with the advice that I offered in my TEDx talk. I espoused the idea of "living mathematically, but not by the numbers" as an approach to life that highlights the skills that are important for solving math problems, but applied to everyday life. There are four qualities that I think we should try to enhance in ourselves: creativity, flexibility, confidence and determination, and rigor. Let me explain.

1. Creative people find ways to handle life's challenges. I believe that curious and creative people tend to be happier people. Their open minds allow them to appreciate the world's infinite complexity.
2. Flexible people are able to approach problems from different points of view. Flexible people are equipped to meet and work with people from different backgrounds and cultures.

3. Confident and determined people have the courage and strength to manipulate difficult ideas. They have the confidence to attack challenges even when a solution is not obvious.

4. Rigorous people pay attention to details, and they are more likely to find opportunities in unexpected places.

Live mathematically, but not by the numbers.

Chapter 34

∿

MY SPIRITUALITY

I am often asked whether I believe that Ramanujan's findings truly came to him as visions from a goddess. I didn't believe this for most of my life. Perhaps if I were a Hindu, I would have had an easier time subscribing to such a view. Instead, I wish to offer my opinion on a simpler question: was Ramanujan's mathematics divine in origin?

I have read most of Ramanujan's papers multiple times. I have read virtually everything ever written about him, and I have read and reread his letters and notebooks many times. I have tried to develop some sense of what he was like, and what motivated his thinking. The deeper I dig, the more in awe I am of Ramanujan. Because of my growing sense of wonder, I have thought quite a bit about the source of Ramanujan's ideas.

Earlier, I was convinced that his claim of visions from a goddess was poppy-cock. But now I have changed my mind. His claims and formulas, as intimidating as they are at first glance, are awe-inspiring in their beauty and rightness. The more I read Ramanujan's work, the greater are the depths that are revealed. How was it possible for an untrained youth ignorant of modern mathematics to produce those wonderful formulas? Reading Ramanujan's writings has become a spiritual experience for me. I sense in his revelations to me a divine source of revelation to him.

I now firmly believe that Ramanujan's ideas are divine in origin, though I am much less sure just what I mean when I use the word "divine." I am not saying that Ramanujan had a direct line to God, whatever that might mean. Instead, I share Carl Sagan's view that "science invariably elicits a sense of reverence and awe ... The cumulative worldwide build-up of knowledge over time converts science into something ... that is surely spiritual." From this point of view, I believe that all science is spiritual. Francis Collins, director of the National Institutes of Health and the former leader of the project that mapped the human

© Springer International Publishing Switzerland 2016
K. Ono, A.D. Aczel, *My Search for Ramanujan*, DOI 10.1007/978-3-319-25568-2_34

genome, is a strong advocate of this view, and he is well known for having said that "The God of the Bible is also the God of the genome. He can be worshiped in the cathedral or the laboratory."

So I leave aside the question of the nature of God. Our knowledge of the divine can be obscure at best. As Saint Paul tells us, "now we see through a glass, darkly." But like Carl Sagan, like Francis Collins, like Alfred Tennyson in his "Higher Pantheism" that was quoted earlier in this book, I see evidence of the divine everywhere. And while some may find it in the cosmos, or the genome, or the seas, the hills, and the plains, I have seen it most vividly in the work of Ramanujan. It doesn't matter whether Ramanujan believed in the literal existence of the goddess Namagiri or whether he saw in her merely the form that divine inspiration took in his sleeping mind. A version of Namagiri that Western readers may find more accessible can be found in Coleridge's poem about Kubla Khan and his famous pleasure dome, in which the lyric voice records a vision, not of a goddess but of a damsel with a dulcimer, and tells us that a creative person's task is to convert such visions into poetry, or art, or mathematics:

> *Could I revive within me*
> *Her symphony and song,*
> *To such a deep delight 'twould win me,*
> *That with music loud and long,*
> *I would build that dome in air,*
> *That sunny dome! those caves of ice!*

And such a person is indeed divinely blessed:

> *For he on honey-dew hath fed,*
> *And drunk the milk of Paradise.*

My search for Ramanujan has transformed me. Raised with no religion, I found the strong need to come to terms with the sense of awe elicited by Ramanujan's formulas, and more generally the beautiful infinite complexity of the universe. Thanks to this awakening, I now wish to experience the world, to enjoy its beauty and its people. My curiosity pushes me, and it has left me open to discovering my spirituality.

On my thirty-fifth birthday, March 20, 2003, the United States, in a coalition with British, Australian, Polish, and Danish military forces, invaded Iraq with the goal of toppling the regime of Saddam Hussein. Despite the precision of modern military technology, there were many innocent victims at the outset, Iraqi men, women, and children accidentally killed in the line of fire. Within days of the invasion, a church in Madison, Wisconsin, home to the University of Wisconsin, where I was a professor, began planting white flags on their lawn to honor those children. It was a church that we had driven by countless times without taking notice. With the flags, however, the church got our attention, and as the conflict dragged on, the lawn of the church was transformed into a sea of white flags, a poignant expression of compassion and humanity in memory of those innocent victims. That garden of flags, a symbol of the congregation's prayer for a quick end to the hostilities, provided a daily reminder of our good fortune in a world in which so many others were victims of conflict.

Impressed by this moving display, Erika and I decided to attend a Sunday service. Although religion had played little role in our family up to that point, apart from attending services in Erika's church of origin whenever we were in Montana, we felt a strong need to meet this congregation. It turned out that the church, Madison Christian Community, is a partnership between two congregations: Community of Hope and Advent Lutheran Church. Although the two congregations have different religious affiliations, they are partners. They have common educational and community outreach activities; they even have a joint church service once a month.

We attended a service at the Community of Hope, an open and affirming congregation affiliated with the United Church of Christ. We were drawn by their statement of welcome:

Community of Hope is an Open and Affirming community of faith
where no matter who you are or where you are on life's journey,
you are welcome.
We come together
As people on a journey,
Respectful of our diversity,
Supportive of our callings and committed to learning what it means
to be the body of Christ in the world today.
We celebrate all races, abilities, creeds, socioeconomic statuses, religious
upbringings and sexual orientations.

Having grown up without religion, I was comforted by those words. We were indeed welcomed and embraced by this community on our first visit, and we realized on the spot that we had been leading lives that lacked spirituality and a sense of community.

The congregation was led by Pastor Tisha Brown, an intelligent and vibrant woman whose weekly sermons erased my earlier fears of church: How would the congregation view heaven and hell? Was it really important to choose the right version of the Bible? How should Christians be viewed within the context of a worldwide community? How would other religious traditions be viewed? None of those questions mattered. The congregation that Tisha led was indeed open and affirming in every way.

Tisha challenged us to deepen our spirituality and to perform work in search of justice and an improved quality of life for people everywhere. For this former tiger boy who was raised in a nonreligious kaikin home, her messages shook the foundations of my very being. Her sermons and the friends we made at the Community of Hope challenged me to rethink my purpose; they helped me develop my spirituality; and they helped me become a better person.

In 2004, I asked Tisha to baptize me. As part of the process, I had to meet with her to discuss the church, the community, and my expected role as a future member. As a prerequisite, she asked me to explain my personal spiritual path. I came to our first meeting prepared. I brought Kanigel's *The Man Who Knew Infinity* with me, and I told her about my life and the great Indian genius whose ideas came to him as visions from a goddess.

What would it take for my parents to open themselves to the possibility of a life of the spirit? I asked at the beginning of this book. Recall that during their childhood, Emperor Hirohito, who was viewed as a god to the Japanese people, had been forced to surrender and abandon his divine status. I theorized that this event left many Japanese, like my parents, jaded and closed to any form of spirituality.

My parents now belong to a church in Lutherville. For the past ten years, they have enjoyed their membership in a spiritual community. Although we have never talked about it, I believe that my father, like me, has learned to recognize and celebrate the things of this world and of the mind and spirit as fragments of the divine. I had run away from home so that I could be as little like my father as possible. I was surprised when Basil Gordon prophesied that I would follow in my father's footsteps. I was even more surprised when Ramanujan emerged as the link that bound my father's history to mine. Did he also have a role in bringing my parents to think more about spirituality and community?

EPILOGUE

My Pilgrimages

P ilgrimages are not unique to one religion. They are retreats from normal life to focus on spiritual values or to honor a particular place or person. Because of my life story, I had a strong need to make a pilgrimage to honor Ramanujan.

Almost fifteen years ago, around the time I moved to the University of Wisconsin from Penn State, the Dutch mathematician Sander Zwegers, in his doctoral dissertation written under the direction of Don Zagier, finally made sense out of Ramanujan's mock theta functions, the mathematics he had conjured on his deathbed. Almost concurrently, the German mathematicians Jan Bruinier and Jens Funke developed a general theory of such functions that then made use of Zwegers's work to show that Ramanujan had anticipated the theory of "harmonic Maass forms."

Armed with these advances, Kathrin Bringmann and I proved a number of theorems on Ramanujan's mathematics that earned us many invitations, one of which came from India. I was invited to speak at the 2005 SASTRA University conference on number theory. SASTRA University, located in Ramanujan's hometown of Kumbakonam, had decided to establish an award in honor of Ramanujan, a prize to be bestowed on mathematicians not exceeding the age of thirty-two who have made outstanding contributions to areas of mathematics influenced by Ramanujan. I was invited by SASTRA to give an address at the inaugural SASTRA Prize Conference.

My dream had finally come true. I was offered the opportunity to make a pilgrimage to India, to seek Ramanujan, to see his home, his temple, and his schools. Krishnaswami (Krishna) Alladi, a number theorist at the University of

© Springer International Publishing Switzerland 2016
K. Ono, A.D. Aczel, *My Search for Ramanujan*, DOI 10.1007/978-3-319-25568-2

Florida and the founding editor of the *Ramanujan Journal*, kindly facilitated our visits to these special places. My travel companions were the SASTRA award winners, my friends Sound and Manjul Bhargava. This would be my first of many trips to India.

After final exams at the University of Wisconsin in December 2005, I boarded a plane at the Dane County airport. Forty hours later, I was in India, in Chennai, the city where Ramanujan had written his deathbed letter on mock theta functions eighty-five years earlier. This is the city where he passed away tragically at the age of thirty-two with Janaki by his side.

Chennai was not the final destination. I still had the long journey to Kumbakonam, Ramanujan's childhood home and the site of the SASTRA conference. SASTRA University arranged a private minivan, barely large enough to accommodate the American mathematicians attending the conference, which included Sound, Manjul, Manjul's mother, Krishna, and my doctoral student Karl Mahlburg.

My arrival in Chennai had been delayed by a winter storm in Europe, which left almost no time to relax before the long bus ride. Manjul graciously offered his hotel room for a quick shower, after which we departed for Kumbakonam straightaway.

Although Kumbakonam is only 180 miles from Chennai, the drive took over six hours. At first, we poked along in the ridiculous congestion that defines Chennai traffic. Imagine inching along in a sea of bicycles, small cars, cows, goats, motorcycles, and rickshaws—stoplights being mere suggestions. I assumed that the traffic conditions would improve once we escaped the city. In some ways, they did. Our speed improved outside the city, but the driving conditions deteriorated. The roads narrowed, and they had been severely damaged by recent flooding, which had scoured out deep potholes. That preposterous ride rivaled the most extreme turbulence I have ever experienced in an airplane, and it went on and on and on for nearly thirty miles. I had no idea that a 180-mile drive could be so physically demanding. We arrived in Kumbakonam after dusk, and we were greeted by the staff of the hotel, who draped lovely garlands around our necks and imprinted red tilaks on our foreheads. They kindly offered glasses of delicious rose water, perfect elixirs after such a grueling ride, and then they massaged our feet.

Our hotel, the secluded Sterling Resort, was a collection of carefully restored nineteenth-century buildings that had been an abandoned rustic village. The grounds were complete with a small farm, livestock, large sculptures of Hindu gods, and a museum.

The next morning, I took a short stroll around the grounds. It seemed as though we had somehow taken a journey back in time. After a breakfast of masala dosa, one of my favorite south Indian dishes, we boarded the minivan for the short drive to SASTRA University, the site of the "Conference on Number Theory and Mathematical Physics" and home of the Srinivasa Ramanujan Centre. The day began with the presentation of the first SASTRA Ramanujan Prize, awarded jointly to my friends Manjul and Sound. The dazzling ceremony included the lighting of a tall polished brass lamp, traditional Indian songs, and a passionate speech by the executive director of the Indo-US Science and Technology Forum, an organization that promotes collaboration in science, technology, engineering, and biomedical research between the United States and India.

After a full slate of lectures, the invited speakers were driven to two sacred sites: Ramanujan's childhood home and the Sarangapani Temple. We first visited Ramanujan's home, a one-story stucco house that sits inconspicuously among a row of shops. This house, which had deteriorated and was in a dilapidated condition for several decades, was purchased and beautifully restored by SASTRA University. At SASTRA's invitation, India's president, Abdul Kalam, visited the home in 2003, and he was so impressed that he declared it a national museum.

Sound and Manjul in front of Ramanujan's boyhood home (photo by Krishnaswami Alladi)

The house is devoid of any striking features. In the front, there is a small porch, one of Ramanujan's favorite places to do mathematics. We took many photos of the porch, and we tried to imagine Ramanujan as a young boy, performing his calculations there on his slate. I spent the next half hour pacing through the diminutive house, which consisted of two small rooms and a kitchen.

The very small bedroom is found immediately on the left as you enter the front door, and its only distinguished features are a small window facing the street and an old-fashioned bed occupying nearly half the floor space. The museum's exhibits, which include a bust of Ramanujan decorated with garlands, are lovingly displayed in the main room. On the day we visited, there was a beautiful *kolam* in front of the bust, an intricate, symmetric floral design on the brick floor created out of rice flour. In the rear part of Ramanujan's house there is a tiny courtyard with an old well.

Two blocks away, the Sarangapani Temple towered over Ramanujan's neighborhood. There, Ramanujan and his family regularly offered prayers to the Hindu god Vishnu, one of whose epithets is *Sarangapani*, the bow-carrier. Ramanujan used to work on his mathematics in its great halls, sheltered from the heat and humidity by its stone walls. I tried to picture Ramanujan working out his mathematics on his slate with his notebook by his side.

The Sarangapani Temple (photo by Krishnaswami Alladi)

The brilliant orange of the sun's rays formed a corona around the colossal structure, which beckoned us as we stood on the porch of Ramanujan's house. The temple, built mostly between the thirteenth and seventeenth centuries, was constructed from stone brought from north India by elephant. The temple is tetragonal, and its outer walls are completely covered with colorful ornate carvings depicting countless Hindu legends.

Immediately beyond the *gopuram*, the temple gate, dozens of bats circled above us against the darkening sky. A few steps away, we could see cows eating hay. The interior of the temple is a stunning labyrinth of sculptures, stone columns, brass walls, flickering lights and candles, and brass pillars. The walls are completely covered with ornate metalwork and stone carvings. Following Hindu tradition, we stepped barefoot over the stone floor in a clockwise direction, passing dozens of *kolam* floor designs. The air was warm and muggy, and heavy with the scent of incense. The main central shrine is a monolith resembling a chariot drawn by horses and elephants. Beyond the monolith we found the inner sanctum, protected by a pair of ancient bulky wooden doors covered with bells. The inner sanctum, bursting with silver and bronze vessels, is considered the bronze-walled resting place of Lord Vishnu. Krishna and his cheerful wife, Mathura, called us into the inner sanctum and made offerings of coconuts and vegetables to Lord Vishnu, placed by sweaty bare-chested Hindu priests clad in holy white robes. I understood that Krishna had arranged for us to be blessed in this impassioned *pooja*, or ceremony of propitiation.

As we made our way back toward the gate of the temple, I came upon a small set of steps that led to a small cubbyhole holding the statue of a Hindu god flanked by melted candles. Its ancient stone walls were covered with numbers scrawled in charcoal and carved in stone. Ramanujan's temple had special numbers! Sound's father explained that it is not unusual for Hindus to etch important numbers when making offerings. Some numbers appeared to be birthdays, while others appeared to be telephone numbers and street addresses. I excitedly searched for numbers made famous by Ramanujan, such as Hardy's "taxicab" number (to be described later). I didn't find any. However, to my great surprise, I found 2719 prominently etched at eye level. For me, this number is special. It is the largest odd number that—as Sound and I had proved with near certainty—is not represented by Ramanujan's ternary quadratic form $x^2 + y^2 + 10z^2$. I was delighted to see it near where Ramanujan might have been writing in his notebook a century ago. I have now visited India several times, and I always make it a point to check on my sacred 2719 etched delicately in a niche in Ramanujan's temple.

The next day provided another full program. My student Karl gave a superb talk on his thesis research, which would earn him a postdoc at MIT and the *Proceedings of the National Academy of Sciences* "Paper of the Year Award." I then gave my lecture on my work with Kathrin on mock theta functions and Maass forms. Kathrin would win a SASTRA Ramanujan Award a few years later in large part because of these results.

At 3:00 p.m., we boarded the minivan for further sightseeing. We visited Government College, the first college to dismiss Ramanujan, and Town High School, where he excelled before becoming addicted to mathematics. At Government College, I had hoped to see the original copy of Carr's book, the one that awakened Ramanujan's genius. Unfortunately, the book was lost.

After the short visit to Government College, we made our way to Town High School, the site of Ramanujan's first academic successes. We arrived after classes had ended for the day. The school was an impressive two-story building with arched balconies and a lush tropical courtyard. We were greeted by A. Ramamoorthy and S. Krishnamurthy, two of the school's teachers. They kindly gave us an entertaining tour of campus, which included a stop in *Ramanujan Hall*, a cavernous room dedicated to the memory of Ramanujan. The teachers also proudly displayed copies of awards that Ramanujan had won as a top student. I was deeply moved by the pride with which they shared their campus and their devotion to the story of Ramanujan. Their passion confirmed to me that Ramanujan is still treasured in India.

Near the end of our visit, Mr. Ramamoorthy revealed to us that he teaches English, and as a student was never very good at math. He timidly asked whether I could explain any of Ramanujan's work to him, and from the look on his face, it was clear that he didn't think that I could. I accepted the challenge. We found a chalkboard, and I explained Ramanujan's work on partitions. I explained the partition numbers to him, and then told him that Ramanujan proved that every fifth partition number, beginning with $p(4) = 5$, is a multiple of 5. My new friends were delighted by the simplicity of the result, and they promised to share it with their students the next day. Mr. Ramamoorthy thanked me, and he joked, "You are not Mr. Ono, you are Mr. Oh-Yes." Although I would have hated hearing this as a high-schooler, somehow in Kumbakonam it was music to my ears.

The conference ended the next day, on December 22, 2005, Ramanujan's birthday. Manjul delivered the Ramanujan Commemorative Lecture, a fascinating look at his work with Jonathan Hanke on quadratic forms. Manjul fittingly noted that in his 1916 paper on quadratic forms, Ramanujan had already anticipated some of the most difficult problems in the subject. Manjul's final slide was about my joint work with Sound, our result that proved (modulo the generalized

Riemann hypothesis) that 2719 is the largest odd number that is not of the form $x^2 + y^2 + 10z^2$. That slide was for me a poetic conclusion to the conference and my pilgrimage: the number 2719 appearing at the end of the final lecture and echoing from a small cubbyhole in the great halls of Ramanujan's temple a few miles away.

Three years ago, I had the chance to visit many of the other sites that are important to Ramanujan's history. The president of India had proclaimed 2012 a "National Year of Mathematics" to honor the 125th anniversary of the birth of Ramanujan. The Indian government provided the financial backing for the documentary *The Genius of Srinivasa Ramanujan*.

Nandan Kudhyadi, the director, asked me to appear in the film as one of the experts on Ramanujan's work, along with my friend A. Raghuram, the coordinator of the mathematics program at the Indian Institute of Science Education and Research, in Pune, India. I spent several weeks in October 2011 with a small film crew visiting sites in south India that played an important role in Ramanujan's life. I visited Ramanujan's birthplace in Erode. We visited Namakkal, a town that dates to the seventh century. Namakkal is home to the Namagiri Temple where Ramanujan had his vision that allowed him to accept Hardy's invitation and travel across the seas to Cambridge. We filmed at various locations in Madras, including the hostel where Ramanujan lived before flunking out of college and the neighborhood where he spent the last months of his life.

I even held, in my own hands, Ramanujan's notebooks, the ones that have been central to my mathematical career. As I held them, I was thinking, "Oh my God! Is this really okay? Have I earned the right to hold these sacred volumes?" We spent hours at the University of Madras, studying Ramanujan's notebooks. I could have spent days examining its pages. Holding the notebooks, the source of so many treasures, was a spiritual experience that I will never forget. It is one of the highlights of my life as a mathematician.

It was clear that Ramanujan had edited these notebooks carefully—they clearly did not contain the raw results as he first obtained them. His mathematical formulas and equations are carefully written, on neat pages, usually only on the right side of the page, although the left side would later be used as scrap paper for calculations. Ramanujan wrote very carefully in green ink, clearly and legibly.

Ramanujan's notebooks (published after his death and edited by Bruce C. Berndt) contain some work he did he while still a student at Town High in Kumbakonam. Flipping through his first notebook, I immediately came across some of that work, his results on magic squares, which are square arrays of numbers such that the sum of the numbers in each row and column and on the main

Perusing Ramanujan's notebooks with Raghuram (photo courtesy of Emory University)

diagonals, and often in subsquares within the main body of the square, add up to the same number.

Magic squares with three rows and columns, called 3 by 3 (or 3×3) squares, have been known since early antiquity. Chinese writings from about 650 B.C.E., the Lo Shu, or "Scroll of the River Lo," contain a 3×3 magic square, an arrangement of the numbers from 1 to 9:

$$
\begin{array}{ccc}
4 & 9 & 2 \\
3 & 5 & 7. \\
8 & 1 & 6
\end{array}
$$

Note that the sum of every row, of every column, and of the two main diagonals is 15, and that is what characterizes a "regular" 3×3 magic square. For a 4×4 magic square, containing the numbers from 1 to 16, the sums must be 34. In fact, in the tenth-century temple complex in north India called Khajuraho, there is an interesting 4×4 magic square, attesting to the fascination in ancient India for such objects. The fact that this square is in a temple suggests some religious or spiritual connection between mathematics and the gods.

I stared in awe at the huge 8×8 square Ramanujan had constructed and recorded in his first notebook. He also discussed some properties he saw in magic squares, for example, that the middle row, middle column, and each diagonal in a 3 × 3 square form an arithmetic progression. Arithmetic progressions are sequences in which the adjacent numbers share a common difference. In the 3 × 3 square above, the arithmetic progressions in question are (3, 5, 7), (9, 5, 1), (4, 5, 6), and (8, 5, 2).

The documentary provided me with a second pilgrimage to pay homage to Ramanujan. This time, I had the chance to share it with Indian filmmakers and mathematicians who would be telling the story to the public. This trip, with Kudhyadi, Raghuram, and my new friend Venkateswaran Thathamangalam Viswanathan, filled in many of my cultural gaps in the Ramanujan story.

Face to Face with Ramanujan

If anyone were to ask me what one person I would want to meet in all of history, my answer, as the reader may easily guess, would be Srinivasa Ramanujan. Amazingly, my wish came true, in a way.

In the last week of July 2014, I received an unexpected email from film director Matthew Brown. I had heard eight years earlier that Matt had written a screenplay for a film on Ramanujan based on Robert Kanigel's book *The Man Who Knew Infinity*. I hadn't heard anything further about it in years, and so I naturally assumed that the project had been scrapped.

I was delighted to learn from Matt that the film was well into preproduction. I was already excited by the promise of the forthcoming "mathy" biopics *The Theory of Everything*, about Stephen Hawking, and *The Imitation Game*, about Alan Turing. Those films would soon earn wide critical acclaim, including multiple Oscar nominations. Both films would go on to win Academy Awards. Perhaps this film on Ramanujan would be the next British math prestige film.

Dev Patel, the twenty-four-year-old star of the hit *Slumdog Millionaire*, was cast to play Ramanujan, and Academy Award winner Jeremy Irons was cast to play Hardy.

The Pressman Film Company, producers of over eighty films including *Conan the Barbarian*, *The Crow*, and *Wall Street*, was producing the film at Pinewood Studios, in London. Filming was set to begin early in August.

Matt's email was short. He wanted to skype the next morning to discuss the film. He mentioned something about technical assistance for the art department. I had no idea that this would lead to a whirlwind opportunity of a lifetime, the closest I would ever get to meeting Ramanujan face-to-face.

The next morning, I skyped with Matt and Liz Colbert, an Irish graphic artist well known for her work on the TV show *Game of Thrones* and the blockbuster film *Sherlock Holmes*, which starred Robert Downey Jr. Liz was charged with producing the mathematical props for the film. She needed help in identifying the documents to be used, which would have to be painstakingly reproduced by hand. My job was to choose these documents, including some of Ramanujan's letters to Hardy and select pages from the notebooks.

Our skype session extended well beyond the scheduled thirty minutes. There was simply too much to discuss, and much work to do. I could tell that the art department was relieved to know that I knew my stuff. I knew the documents by heart, and I knew the story. Within minutes, I was the man in charge of the math. I was thrilled that Matt had decided to pay attention to such details. There was to be no fudging. He wanted to be true to the story, and that included exceptional attention to detail regarding the mathematics in the film.

Matt thanked me for my offer of help, clearly relieved that getting the math right was no longer a worry. He had no idea that I would have dropped everything to work on the film.

I wanted to know about the last-minute tasks that had to be completed before filming could begin, and I learned about the frenetic last days of preproduction. Matt had to finalize schedules, oversee the production of props, visit locations, and he had yet to finish casting. Among his concerns, he mentioned a little detail that the research department had been struggling to solve. They had been unable to find Janaki's autograph. Even though it probably would not appear in the film, Matt wanted it just in case. Did I have any leads?

I proudly exclaimed that I didn't need any leads, because I had Janaki's autograph. In fact, her autograph had been in my possession for fifteen years, and in my father's for the fifteen years before that, since that day in 1984 when her letter to my father arrived in our mailbox in Lutherville, Maryland. Matt and Liz looked at each other and laughed in surprise at their good luck. Luck? I wasn't so sure.

A few hours later, Matt emailed me again with a handsome offer that I couldn't refuse. So I didn't, and three days later, I found myself on a flight to London. I would end up spending several weeks working on the film. At first, my job was to help the art department with the formulas. I had no idea that I would end up doing much more. In fact, I would ultimately be named an associate producer.

The Pressman Production Company arranged a luxurious hotel room for me near Paddington Station, the location being based on my requirement of proximity to a competition swimming pool. I was training for the International

Triathlon Union's Cross Triathlon World Championships, which would be held in Zittau, Germany, at the end of August.

Preproduction at Pinewood Studios, which was also filming *Star Wars: Episode VII*, was frantic. The art department was busily producing props, including copies of Ramanujan's letters and notebooks. Liz and I carefully went over the documents that she needed scene by scene.

Three days before filming was to begin, Jeremy Irons and Dev Patel arrived at the studio for rehearsals. I had hoped to get a glimpse of them working with Matt. I was working with Liz in her office when a production assistant rushed in with the message that the principals wanted to see me.

A few minutes later, I was in a quiet room with Matt, Jeremy Irons, and Dev Patel, secluded from the rest of the preproduction frenzy. Matt introduced me, and I nervously shook their hands. Matt handed me a script and said that we were going to rehearse. Oh my God! Oh my God! I couldn't believe it. I was going to get more than a glimpse!

Jeremy and Dev had asked for me because they wanted to know how a mathematician thinks. No true actor is content merely to deliver his lines. How did Hardy and Ramanujan think about mathematics? How did they speak about it? Write about it? What sort of gestures would they have used? They explained to me that what they wanted from me was something like what actors get from a dialect coach. They couldn't slip into their roles without an expert on hand explaining how these men would have thought and spoken.

The four of us sat in a circle in a second-floor room at Pinewood Studios. This was the place that had made the James Bond films, the Harry Potter films, the Captain America films, and the Sherlock Holmes films. Oh my God, and I was there with a job to do! This was more than a tour of the Wonka chocolate factory. I was going to help make the chocolate!

Those rehearsals were amazing in many different ways. As a fan of film, it was my chance to observe firsthand some of the world's finest actors at work. I had no idea how difficult it is to be a serious actor. Our work on the scenes followed a standard pattern. Matt would begin by giving the outline of each scene and its role in the film. Then the actors would read the lines out loud. The three of them would then discuss what they had to do to make the scene work. Perfecting a scene involved reworking lines, working out facial expressions and body language, and discussing the props, all in a quest for authenticity.

With Jeremy Irons on the set (photo by Sam Pressman)

They all had lots of questions for me. By the end of the first hour, I understood what was required of me, and I was comfortable enough to speak up and offer suggestions and comments. They trusted my opinion and judgment as the math consultant. All along, I was thinking that my passion for Ramanujan and his mathematics was about to make the world stage, performed by the world's best. Unbelievable!

Going over the script with Dev and Jeremy was beyond a dream come true. It wasn't even something I would ever have dreamt of wishing for. Although I doubt that Hardy ever called anyone "babycakes," as Jeremy did me, I will forever cherish these rehearsals as the closest I will ever come to being face-to-face with Hardy and Ramanujan.

To celebrate the end of preproduction, Edward Pressman hosted a small private party at a fancy London restaurant. To my surprise, he asked me to say a few words about Ramanujan and his mathematics. Although I am accustomed to giving lectures about Ramanujan, this short speech was something special. With a glass of champagne in hand, I did my best to speak eloquently about Ramanujan, the man I had been searching for most of my life. And there was Dev Patel, the man who would play Ramanujan, seated directly across from me.

Preproduction party (*left to right*: Edward Pressman, Jeremy Irons, Dev Patel, Ken Ono, Matt Brown, Sorel Carradine) (photo by Sam Pressman)

My Search Goes On

Building on the success I enjoyed in seeking the hidden meaning in Ramanujan's writings—his claim that there are no other simple properties for partitions other than the ones he had found, his search for a simple law that determines the odd numbers not of the form $x^2 + y^2 + 10z^2$, I decided to study his works in detail on the belief that Ramanujan, speaking beyond the grave, has vouchsafed hints of beautiful theories to the mathematicians of today. I hear him speaking to me when I read his papers. And if I read them with care, I might be lucky enough to figure out what he had in mind.

I have spent the last five years as a professor at Emory University, working on some of Ramanujan's mathematics and its implications for contemporary questions. Jan Bruinier and I revisited the partition numbers and the work of Hardy and Ramanujan that gave the approximate formula for those numbers described below in the section on Euler's partition numbers. In the 1930s, building on the Hardy–Ramanujan formula, which gives good approximations but never the exact values, Hans Rademacher obtained an exact formula. However, his formula had a catch. His description of the partition numbers was as a convergent infinite sum. (Since a partition number is an integer, you could use Rademacher's formula to compute individual partition numbers: you know that you have arrived at the correct value after adding a finite number of terms when you can estimate the sum of the infinite number of remaining terms as being less than a small fraction.) We aimed to find a different conceptual formula—one that involved only finitely many terms.

We reformulated the problem in terms of concepts that we had already been using on related problems. By modifying work that Jan had done with Jens Funke, we produced a formula that gives the exact values of the partition numbers as a finite sum of algebraic numbers (these are relatively well behaved rational or irrational numbers) that are themselves values of a single special function. And in joint work with Drew Sutherland of MIT, we were able to express our formula in a way that could be, and now has been, implemented on a computer.

I have thought very deeply about the mathematics in Ramanujan's first letter to Hardy. The first challenge of that letter was to understand how Ramanujan came up with his $R(q)$ continued fraction, the one that generalizes the golden ratio. Ole Warnaar, a well-known mathematician at the University of Queensland, had been working on this problem too.

I learned two years ago that Ole had made a huge breakthrough concerning certain hypergeometric series transformation laws, results that experts believed could hold the key to finding more identities of the type that Rogers and Ramanujan had discovered a century earlier. Ole had found infinite families of power series identities that included the special continued fractions that Ramanujan discussed in his first letter to Hardy.

I immediately contacted Ole by email, and I suggested that we join forces. Together with my PhD student Michael Griffin, we made use of Ole's beautiful formulas to establish a framework in which to place the Rogers–Ramanujan identities, which we then put to good use to obtain infinitely many generalizations of the golden ratio, algebraic units that are the values of the functions that Ole had discovered. What Ramanujan had offered in his first letter to Hardy turned out to be the first example of the functions we now understood.

Discover magazine ranked our accomplishment fifteenth among the top hundred stories in science of 2014. The editors conducted a "People's Choice Award," and our work on Ramanujan's first letter came in second.

Like his first letter to Hardy, Ramanujan's last also speaks to the mathematicians and physicists of today. In that letter, he described his enigmatic mock theta functions, which have been studied by the Japanese physicists Tohru Eguchi, Hirosi Ooguri, and Yuji Tachikawa and the Canadian mathematician

George Andrews, Michael Griffin, Ken Ono, Ole Warnaar, Jim Lepowsky

Terry Gannon. Building on their work, Miranda Cheng (University of Amsterdam), John Duncan (Case Western), and Jeff Harvey (UChicago) formulated a conjecture in 2010 that predicted a very deep and precise relationship between Ramanujan's mock theta functions—the stuff of Ramanujan's deathbed letter to Hardy—and the hottest item in theoretical physics today: *string theory*.

Their conjecture, which they called the *umbral moonshine conjecture*, was formulated to place Berkeley mathematician Richard Borcherds's 1992 proof of the so-called moonshine conjecture, for which he was awarded the Fields Medal, in a wider context. If correct, Borcherds's work would be the first of many different "moonshine" theories. Mathematical physicists were now realizing that such theories are important in string theory, which aims to answer a fundamental question: what is the universe made of?

In joint work with John Duncan and my PhD student Michael Griffin, we proved the umbral moonshine conjecture, a result that now places the mathematics of Ramanujan's deathbed letter front and center in cutting-edge

With Ramanujan's last letter to Hardy in 2013 at Trinity College

mathematical physics. We had proved that the functions he conjured in the last months of his life encode astonishing symmetries in the world of mathematics, and experts now predict that his functions will be put to good use in the study of black holes, quantum gravity, and the theory of everything. *Discover* magazine has informed us that our accomplishment will be among the top 100 stories in science of 2015.

John Duncan, Ken Ono, Michael Griffin

The search for Ramanujan shall go on. His words still speak to us.

My mathematical search for Ramanujan is a never-ending story. As I continue my quest, I find again and again that I have overestimated my ability to understand the full meaning in Ramanujan's notebooks, letters, and papers. Ramanujan's legacy is inexhaustible.

AFTERWORD

Two Questions

I am often asked about Ramanujan and his story. Was Hardy the best mentor for Ramanujan? Was Ramanujan the greatest mathematician of his time—or perhaps of *all* time?

Was Hardy the best mentor for Ramanujan? This is a difficult question to answer. People have raised this question for a variety of reasons. Hardy was not an expert on theta functions and modular forms, areas in which one arguably finds Ramanujan's most important contributions. Would mathematics have advanced further had he been mentored by an expert in those fields? Some say that Hardy selfishly made use of Ramanujan's intellect instead of helping him to become a more professional mathematician. Would Ramanujan have been a more important mathematician had he been mentored by someone who did not need him as a collaborator?

Although these are legitimate questions, the point is moot. We cannot travel back in time and change history. But if we must wonder, here are my thoughts. In India, Ramanujan worked in a vacuum; there was no one to nurture his talent. Hardy recognized Ramanujan's ability when others ignored him, and he brought him to England and helped him make his mathematics understandable and presentable to the world. Had this never happened, then Ramanujan would likely have disappeared without a trace. Had this never happened, I would never have happened. I will therefore always hold Hardy in very high regard. Opinions vary on whether Hardy could have taken steps that would have saved Ramanujan's life. I am not qualified to speculate, and I don't wish to speculate.

© Springer International Publishing Switzerland 2016
K. Ono, A.D. Aczel, *My Search for Ramanujan*, DOI 10.1007/978-3-319-25568-2

It is true that there were mathematicians who might have recognized Ramanujan's talent and creativity in the theory of modular forms and theta functions. However, I doubt that any of them would have been inspired to calculate partition numbers, a problem that deeply interested Hardy. Had Ramanujan had such a mentor, then analytic number theory would have missed out on one of its most important advances, the development of the "circle method." Moreover, had Hardy and Ramanujan never done their work on partitions, then Rademacher would have had nothing to perfect, and the theory of Rademacher sums wouldn't exist. And without the theory of Rademacher sums, much of the current work in mathematical physics wouldn't exist.

The questions that I really wonder about are these: What would Ramanujan have discovered with the help of a computer? What would he have discovered had he not died tragically at the age of thirty-two?

Alas, the world will never know.

Was Ramanujan the greatest mathematician of his time—or perhaps *all* time? My answer is best framed in the context of the lovely article "The two cultures of mathematics," by Timothy Gowers, a recent British Fields medalist and current professor and Fellow of Trinity College. In that article, Gowers argues that there are really two kinds of mathematicians, generally speaking: those that are "problem solvers" and those that are "theory builders." Gowers gives two examples to which he thought most mathematicians could relate. The first is Paul Erdős, who with no real home, famously traveled the world, living from his suitcase, working with other mathematicians. Erdős was a problem solver, and his many papers, written with over five hundred collaborators, are solutions to problems, stated, of course, as theorems or other mathematical results. Erdős was so prolific, and published so many papers with so many collaborators around the world, that his colleagues honored him by assigning an "Erdős number" to mathematicians based on the length of the chain of collaboration. Those who collaborated directly with Erdős have Erdős number 1; those who did not collaborate directly but collaborated with those collaborators have Erdős number 2, and so on, in a sort of "six degrees of separation" math game.

As for the other type of mathematician, Gowers gives the example of Sir Michael Atiyah, of Oxford. It is not that Atiyah didn't solve any problems—of course he did, and the famous Atiyah–Singer theorem named for him could be viewed as a problem that the two mathematicians solved. But Atiyah's purpose as a mathematician, as Gowers shows from articles and interviews, was to pursue a theory—to stretch and test mathematics to its limits and then to extend those limits to build more mathematical structures. To be sure, there is an overlap, and theory builders also solve problems, and problem solvers, through their solutions, also extend our knowledge as a whole, and hence build theory. But based on their entire output, Atiyah and Erdős can serve as representatives of the two groups.

Alexander Grothendieck is probably the best recent example of a theory builder. With his students, he rebuilt the field of algebraic geometry. He cared so little for details that he was famously quoted as once saying, when someone interrupted a talk he was giving asking for an example of a prime number, "Well, take 57." Of course, 57 is not prime, since it equals 3×19. Andrew Wiles, of course, could be called a master problem solver for solving the greatest unsolved problem of all time: Fermat's last theorem. But in contrast to the problems solved by Erdős, a behemoth of a theory underlies Wiles's solution, namely the Langlands program, which in a sense is unifying major branches of mathematics.

Ramanujan could well be viewed as a problem solver, because much of his work consisted in actually solving a huge number of extremely complicated problems in mathematics. But there are theories lurking behind those problems, and the problems are part of the theory. Therefore, in Ramanujan's case we should resist classifying him as either a problem solver or a theory builder.

In his biography of Ramanujan, Kanigel uses the following analogy to describe Ramanujan's prowess as a mathematician:

A car mechanic reliant on mechanical instinct may "know" how an engine works yet be unable to set down the physical and chemical principles governing it.

Because of his lack of formal training and the fact that he was set in his mathematical ways by the time he reached Cambridge, Ramanujan could not be a theory builder, simply because he didn't know enough about how mathematics worked as an edifice. But his intuition was supreme—100 on a scale of 0 to 100 that Hardy had suggested, in which he gave himself a mere 25, and the great David Hilbert an 80. His immensely powerful intuition, then, allowed Ramanujan to propose and sometimes prove very difficult and unexpected problems in mathematics, but he was not a Grothendieck—someone who built theories with little concern for individual problems.

So here is my answer to the question whether Ramanujan was the greatest mathematician of his time—perhaps of all time. Ramanujan was a gift; he should be remembered as the greatest anticipator of mathematics. Although he was recognized during his lifetime, his most important ideas, those that have powered mathematicians after his death, were largely viewed as insignificant while he was alive. Those ideas are typically found in his notebooks, letters, and papers as innocent-looking formulas and expressions. Those gems have offered visions of the future, hints of subjects conjured long after his death. For mathematicians, theorems and proofs are works of art, and these formulas are reminiscent of the masterpieces by the Dutch artist Vincent van Gogh, who died at the age of thirty-seven, before his work was fully appreciated.

Ramanujan's formulas have inspired many influential mathematicians, such as Andrews, Deligne, Dyson, Selberg, Serre, Weil, to name just a few, and they have supplied prototypes of deeper objects in algebraic number theory, combinatorics, and physics. Ramanujan was an incredibly great mathematician, certainly up there with the greatest in history. He had a gift of imagination the like of which the world of mathematics had never seen before.

Fermat's Last Theorem and the Tokyo–Nikko Conference

The 1955 Tokyo–Nikko meeting, which was a pivotal event in my father's life, turned out to be a pivotal event in the history of mathematics. At that conference, Yutaka Taniyama, one of my father's close friends, posed a problem that suggested a deep connection between seemingly unrelated objects: "modular forms" and "elliptic curves." This problem would evolve over time, and its final form became known as the Taniyama–Weil conjecture, the Shimura–Taniyama–Weil conjecture, and the modularity conjecture. Goro Shimura, who also attended the symposium, was a star, a leader among the young ambitious Japanese mathematicians.

Although number theorists understood the importance of the modularity conjecture early on, it was the work of Berkeley mathematician Ken Ribet in the late 1980s that catapulted the conjecture to prominence. Ribet proved that the conjecture implies Fermat's last theorem.

Ribet's work provided a new approach to Fermat: prove the modularity conjecture. Although mathematicians were excited to learn of the deep connection between the modularity conjecture and Fermat's last theorem, few believed that it would lead anywhere. The result was viewed as further evidence that both problems would be difficult to solve, that they might remain unresolved for generations. But then in 1993, Andrew Wiles made world news when he announced that he had proved Fermat's last theorem. That proof made use of Ribet's theorem.

Ken Ribet, Peter Sarnak, Ken Ono in 2014 (photo by Ling Long)

In this way, the proof of Fermat's last theorem, one of the most important events in the history of mathematics, was born from humble beginnings: an innocent question raised by an unproven young Japanese mathematician at a conference intended to promote reconciliation and world peace in the name of science.

Tragically, Taniyama did not live to enjoy the glorious work that sprouted from his conjecture. Suffering from depression and a loss of confidence, he committed suicide on November 17, 1958. One month later, his fiancée, Misako Suzuki, also committed suicide, and she left behind a note in which she wrote, "We promised each other that no matter where we went, we would never be separated. Now that he is gone, I must go too in order to join him."

Their deaths are tragic, but in a way, their suicides may have made sense to their families and friends. In Japanese culture, there is a long history and tradition of suicide. There is the concept of an honorable suicide, such as *seppuku* practiced by the samurai; the crashing of one's plane into enemy ships as practiced by World War II *kamikaze* pilots; and suicide intended to prevent shame from befalling one's family. And Misako Suzuki's suicide is part of a long tradition in Japan of *shinjū*, or love-suicide, the unhappy lovers believing that they will be reunited in heaven.

Mathematical Gems

This book would not be complete without a discussion of some of Ramanujan's works. Here we include a small sample of Ramanujan's mathematical ideas and conjectures.

The Hardy–Ramanujan Taxicab Number

While Ramanujan was lying ill, this time at a hospital in Putney, just outside London, Hardy would come to see him by taking a train from Cambridge to London, and then a taxi to Putney. And so one day, Hardy made the trip to see Ramanujan. In an attempt to lift the ailing mathematician's spirits, Hardy led off with a casual comment about a number, numbers being Ramanujan's favorite topic: "I came here in a taxi with a very dull number: 1729." But to Ramanujan, who was said to be a friend of every integer, that number wasn't dull at all. To Hardy's surprise, Ramanujan gathered what strength he had, jumped up in bed, and cried, "No, Hardy—it is a very interesting number! It is the smallest number expressible as the sum of two cubes in two different ways!" ($1729 = 10^3 + 9^3 = 1^3 + 12^3$). From this event, the mathematical study of *taxicab numbers*—the smallest numbers that can be expressed as the sum of two (positive) cubes in n distinct ways—emerged. To date, only six taxicab numbers have been identified.

Ramanujan was aware of this property of 1729 because of work he had done on a problem studied by Euler that can be found in his notebooks. The number 1729 appears in Ramanujan's works in yet another context, this time related to Fermat's last theorem. It appears that he was thinking about near misses to Fermat's claim. That is, the Fermat conjecture would be false if there existed a positive integer n (greater than 2) and nonzero integers x, y, and z such that $x^n + y^n = z^n$. It turns out that Fermat's last theorem comes "close" to being false— one unit short—for $n = 3$. The number 1729 is the sum of two cubes (powers of $n = 3$): $9^3 + 10^3$, in one of the two possibilities. But 1729 is not itself a cube. If it were, Fermat's theorem would be false. But 1729 "misses" being a cube by only one unit, since 1728 is a cube: $1728 = 12^3$.

Two years ago, I visited the Wren Library at Trinity College, where I studied the Ramanujan archives. To my surprise, I found a page in Ramanujan's handwriting in which he lists families of near misses to Fermat's last theorem, and one he offers is $1 + 12^3 = 9^3 + 10^3$. It turns out that he had discovered identities that mathematicians would later find to be important in algebraic geometry and mathematical physics. Together with my PhD student Sarah Trebat-Leder, we have discovered that these identities can be reformulated as statements about *K3 surfaces* and ranks of elliptic curves, two important subjects that did not exist in Ramanujan's day.

If

(i) $\dfrac{1+53x+9x^2}{1-82x-82x^2+x^3} = a_0 + a_1 x + a_2 x^2 + a_3 x^3 + \cdots$

or $\dfrac{\alpha_0}{x} + \dfrac{\alpha_1}{x^2} + \dfrac{\alpha_2}{x^3} + \cdots$

(ii) $\dfrac{2-26x-12x^2}{1-82x-82x^2+x^3} = b_0 + b_1 x + b_2 x^2 + b_3 x^3 + \cdots$

or $\dfrac{\beta_0}{x} + \dfrac{\beta_1}{x^2} + \dfrac{\beta_2}{x^3} + \cdots$

(iii) $\dfrac{2+8x-10x^2}{1-82x-82x^2+x^3} = c_0 + c_1 x + c_2 x^2 + c_3 x^3 + \cdots$

or $\dfrac{\gamma_0}{x} + \dfrac{\gamma_1}{x^2} + \dfrac{\gamma_2}{x^3} + \cdots$

then

$$a_n^3 + b_n^3 = c_n^3 + (-1)^n \left.\vphantom{\begin{array}{c}a\\b\end{array}}\right\}$$
$$\alpha_n^3 + \beta_n^3 = \gamma_n^3 + (-1)^n$$

Examples

$$135^3 + 138^3 = 172^3 - 1$$
$$11161^3 + 11468^3 = 14258^3 + 1$$
$$791^3 + 812^3 = 1010^3 - 1$$
$$9^3 + 10^3 = 12^3 + 1$$
$$6^3 + 8^3 = 9^3 - 1$$

Ramanujan's page on sums of two cubes and near misses to the cubic Fermat equation (photo courtesy of Trinity College)

Ramanujan derived so many formulas and theorems in Cambridge that it has taken mathematicians decades to resolve his findings. This book would not be complete without discussing some of his results. Here we offer a glimpse of some of his most well known and influential works.

Approximations to π

Early in their collaboration, Hardy and Ramanujan went over the claims that so densely filled his letters and the notebooks that Ramanujan had brought to England. In all, there were three or four *thousand* such results!—filling pages and pages.

The year Ramanujan arrived in England, his first paper was published in an English journal, the *Quarterly Journal of Mathematics*. It was titled "Modular Equations and Approximations to Pi," and it included amazing formulas for

$$\pi = 3.1415926\ldots,$$

well known to every schoolchild as the ratio of the circumference of a circle to its diameter. The number π is an example of an *irrational number*, a number that cannot be described as a simple ratio of integers. One consequence of its irrationality is that its decimal expansion is infinite, running on forever without a discernible pattern, in contrast to the pattern that can be seen in the decimal expansion of the rational number

$$5/7 = 0.714285714285714285714285714285714\ldots.$$

Amateur and professional mathematicians alike have been captivated by the decimal expansion of π. Despite the fact that π is easily described as a simple ratio in a circle, it is not so simple to calculate its decimal places.

I find it incredible that in 2013, ninety-nine years after Ramanujan wrote his first paper with Hardy, the Russian-American brothers David and Gregory Chudnovsky used a home-built supercomputer running a variant of Ramanujan's formula to determine the first 12.1 trillion digits of this number (with earlier estimates of ten trillion digits of π in 2011 and five million digits in 2010). Their algorithm uses a very rapidly converging infinite series that is a relative of Ramanujan's original formula:

$$\frac{1}{\pi} = \frac{\sqrt{8}}{9801} \sum_{n=0}^{\infty} \frac{(4n)!}{(n!)^4} \frac{\left[1103 + 26390n\right]}{396^{4n}}.$$

This work opened the way to very similar, yet more and more accurate, ways of obtaining more and more digits of π.

Highly Composite Numbers

During his second year in England, Ramanujan produced many more new and original papers—some with Hardy, some that were all his own. As Hardy described it, Ramanujan's "flow of original ideas showed no symptom of abatement." Ramanujan worked on the distribution of prime numbers, the Riemann hypothesis, prime factorization of integers, and more. In 1915, Ramanujan defined a new concept: a *highly composite number*.

Integers are the basic building blocks of mathematics, and prime numbers are the basic building blocks of the integers, for every integer can be broken down, multiplicatively, uniquely as a product of primes. For example, $6 = 2 \times 3$ and $15 = 3 \times 5$ and $3,045,684 = 2 \times 2 \times 3 \times 353 \times 719$. Using this multiplicativity, one can list all the divisors of a given number. The only divisor of 1 is 1. The number 2 has two divisors, namely 1 and 2. The number 12 has six divisors, namely 1, 2, 3, 4, 6, 12.

A *highly composite number* is one that has more divisors than any smaller positive integer. The first few highly composite numbers are 1, 2, 4, 6, 12, 24, 36, 48, 60, 120. Incidentally, one theory in the history of mathematics, proposed by the Austrian-American scholar Otto Neugebauer, is that the Babylonians chose base 60 for their number system because a base with a large number of divisors makes it easier to do arithmetic. Ramanujan defined and studied these highly composite numbers, and he studied their frequency.

Euler's Partition Numbers

Ramanujan was a pioneer in a subject that is very dear to me, that of looking at the integers as sums rather than products. The *partition numbers* are the numbers we discussed earlier that show the number of ways in which an integer can be broken down as a sum. To recapitulate from an earlier chapter: The equalities $3 = 2 + 1 = 1 + 1 + 1$ illustrate that there are three ways of "partitioning" the number 3. Next we observe that $4 = 3 + 1 = 2 + 2 = 2 + 1 + 1 = 1 + 1 + 1 + 1$, which shows that there are five ways of partitioning the number 4. Repeating this process of adding and counting for every number n defines the partition function $p(n)$. Thus our examples can be denoted by $p(3) = 3$ and $p(4) = 5$.

As we mentioned before, the partition numbers grow at an astonishing rate. You might not think so from calculating $p(10) = 42$, $p(20) = 627$, and $p(30) = 5604$. But there are nearly four trillion ways of partitioning 200. Obviously, it would be crazy to attempt to list all the partitions of 200 and then count them one by one.

And even if you could, by the time you get to 1000, you would have to deal with $p(1000) \approx 3.61673 \times 10^{106}$. Compare this with the total number of atoms in the known universe, which is "only"—at the highest estimate—4×10^{81}.

There must be a better way to determine such numbers.

The great Leonhard Euler studied this problem in the eighteenth century, and he found a clever way of computing partition numbers that avoids the impossible task of counting the partitions one by one. He found a "recurrence relation," a procedure that computes these numbers in order. For example, his method makes it possible to compute $p(200)$ if one has prior knowledge of the numbers $p(0) = 1$, $p(1) = 1$, $p(2) = 2, \ldots, p(199)$. Euler's procedure was a major improvement, but it is quite cumbersome.

Ramanujan wanted a better result: he wanted a *formula* for the partition numbers that would merely require plugging in for n to get $p(n)$. Working with Hardy, Ramanujan came close to finding such a formula. Hardy and Ramanujan invented the "circle method," a device in analytic number theory (a branch of number theory that uses methods from mathematical analysis) that has become one of the most important tools in mathematics, to obtain an amazing approximation for the partition numbers. They proved the "asymptotic formula"

$$p(n) \sim \frac{1}{4n\sqrt{3}} \cdot e^{\pi \sqrt{2n/3}},$$

into which one can plug in for n and get back a number that is reasonably close to the partition number $p(n)$. Here e is Euler's number $e = 2.718\ldots$. This formula predicts that there should be 199,280,895 ways of partitioning 100. It turns out that the partition number for 100 is actually 190,569,292, which is less than five percent below the approximate value. The Hardy–Ramanujan formula for 500 gives a prediction that is off by less than two percent, and by 5000, their formula is off by less than one percent. Although the asymptotic formula never gives the exact answer, the percentage error shrinks quickly for larger and larger numbers. After Hardy and Ramanujan produced this astonishing approximate formula, it would be many years before significant progress would be made on the problem of computing the partition numbers exactly.

The important innovation that Hardy and Ramanujan brought to this study was a classic example of the field the two of them were championing: analytic number theory. It used a very powerful "continuous" method, Cauchy's theorem, to attack a discrete problem having to do with counting.

This work drew the attention of a particularly intriguing mathematician at Trinity named Percy Alexander MacMahon (1854–1929). MacMahon was known to everyone as "Major MacMahon." The reason for this was that in the decade before Ramanujan was born, the 1870s, MacMahon had a brilliant military career. After graduating from the Royal Military Academy at Woolwich, MacMahon was sent as a military officer to India, and in an interesting coincidence, served in Madras. By the time Ramanujan was born, MacMahon was back in England, first as a military officer, but later also as a mathematician, working in the area of combinatorics.

He did so well in mathematics that in 1890, he was elected a Fellow of the Royal Society (FRS), and eight years later, he retired from the military altogether to devote himself fully to mathematics. He became very interested in partitions. And he had an amazing skill: like John von Neumann many years later, MacMahon was a human "calculator." He could perform very complicated calculations rapidly in his head.

When he met Ramanujan at Cambridge and became aware of the Hardy–Ramanujan work on partitions of numbers, Major MacMahon offered his considerable calculating abilities to help the pair. He would simply compute the number $p(n)$ for successively larger values of n, compiling large tables of exact values against which Ramanujan and Hardy could compare their own numbers, which they obtained from successively improved estimates of the function that they believed would lead to the "true" $p(n)$.

Ramanujan also proved the very surprising divisibility properties discussed earlier. Recall that he discovered that there exists an arithmetic progression all of whose partition numbers are divisible by 5, that is, that $p(4+5n)$ is always a multiple of 5. The sequence of these partition numbers begins $p(4) = 5$, $p(9) = 30$, $p(14) = 135$, $p(19) = 490$, $p(24) = 1575$, all multiples of 5. He also proved, as discussed earlier, a similar striking theorem for 7 and 11, namely that $p(7n+5)$ is always a multiple of 7, and $p(11n+6)$ is always a multiple of 11. These statements are now known as "Ramanujan's partition congruences."

Hardy and Wright state in their textbook on number theory how those arithmetic properties of the partition counter $p(n)$ were derived by Ramanujan:

> *Examining MacMahon's table of $p(n)$, [he] was led first to conjecture, and then to prove, these striking arithmetic properties associated with the moduli 5, 7, and 11.*

Hardy and Wright then present Ramanujan's brilliant proofs of these theorems, all of which exploit infinite series.

The Rogers–Ramanujan Identities

In Cambridge, Ramanujan returned to some of his earlier work from his days in India. He revisited identities that he knew were related to the last formulas in his first letter to Hardy, the ones that Hardy said had defeated him completely. These are the formulas that Hardy believed had to be true because "if they weren't, nobody would have had the imagination to invent them." These formulas are close to my heart; their present-day formulation is

$$1+\frac{q}{(1-q)}+\frac{q^4}{(1-q)(1-q^2)}+\cdots=\frac{1}{(1-q)(1-q^4)(1-q^6)(1-q^9)\cdots},$$

$$1+\frac{q^2}{(1-q)}+\frac{q^6}{(1-q)(1-q^2)}+\cdots=\frac{1}{(1-q^2)(1-q^3)(1-q^7)(1-q^8)\cdots}.$$

While he was seeking to prove these identities, Ramanujan discovered that someone else had already discovered them. It was a mathematician by the name of Leonard James Rogers (1862–1933), at the University of Leeds, who also had the prestigious distinction of being a Fellow of the Royal Society, but had the unfortunate knack for having his work ignored by the rest of the world.

It was in 1917 that Ramanujan discovered, in an 1894 issue of the *Proceedings of the London Mathematical Society*, where it had gone unnoticed for twenty-three years, Rogers's paper on the exact same identities that he had found! Ramanujan contacted Rogers, presumably with Hardy's help, and the two men then worked together and provided a new, joint proof of these relations, which became known as the Rogers–Ramanujan identities.

Unbeknownst to them, the mathematician Issai Schur (1875–1941), an expert in the mathematical area called representation theory working in Berlin, proved these same identities in 1917. The Rogers–Ramanujan proof would appear only two years later, in 1919.

What each of the two identities does is to give us a function, whose variable is q, represented (for each equation) in two different ways. On the left is an infinite sum of fractions involving the variable q, and on the right is the reciprocal of an infinite product of terms involving q. Even Hardy, in his book on number theory with Wright, notes that the proofs are not simple. But we can say something about what these identities mean. Each side of the equation can be expanded as a formal *power series*. To begin with, each expression in the identity is a slight variation of the function $f(x) = 1/(1-x)$, which we can express as a formal power series as follows::

$$f(x) = \frac{1}{1-x} = 1 + x + x^2 + x^3 + x^4 + \cdots.$$

This is the standard "geometric series" that one encounters in calculus. We can now, for example, express the left-hand side of the first identity above as follows:

$$1 + \frac{q}{(1-q)} + \frac{q^4}{(1-q)(1-q^2)} + \cdots = 1 + q(1 + q + q^2 + \cdots) + q^4(1 + q + q^2 + \cdots)(1 + q^2 + q^4 + \cdots) + \cdots$$
$$= 1 + q + q^2 + q^3 + 2q^4 + 2q^5 + 3q^6 + \cdots.$$

And expanding the right-hand side gives exactly the same terms! The fact that the terms match up is *evidence* of a theorem, but it is not a *proof*: how do we know that *all* the terms match, right out to infinity? The actual proof in this case requires deep ideas that involve some tricky manipulation of power series.

These two identities were the secret behind Ramanujan's stunning expressions, which we discussed earlier in connection with the golden ratio ϕ, which had so astonished Hardy.

Ramanujan's "imagination," the like of which nobody else possessed, was that he figured out how to use these two identities together to magically produce examples of numbers like the golden ratio ϕ by choosing for the variable q the "crazy" numbers $-e^{-\pi}$ and $e^{-2\pi}$ in his continued fraction

$$R(q) = \sqrt[5]{q} \cdot \cfrac{1}{1 + \cfrac{q}{1 + \cfrac{q^2}{1 + \cfrac{q^3}{1 + \cfrac{q^4}{\ddots}}}}}.$$

Hardy didn't know, indeed nobody knew, how Ramanujan had done it.

Not only did Ramanujan figure out these numbers, it turned out that his expressions involved only "algebraic numbers," that is, numbers that are roots of polynomial equations with integer coefficients. This is an important class of numbers in mathematics. It includes the golden ratio ϕ, since it is a solution to $x^2 - x - 1 = 0$. In some sense, "most" numbers are not algebraic. The numbers e and π, for example, are not. Since algebraic numbers are relatively rare, it is amazing that the two wild examples from Ramanujan's first letter that defeated Hardy completely,

$$R\left(-e^{-\pi}\right) = e^{\frac{\pi}{5}}\left(\sqrt{3-\phi}+1-\phi\right),$$

$$R\left(e^{-2\pi}\right) = e^{\frac{2\pi}{5}}\left(\sqrt{2+\phi}-\phi\right),$$

correspond to solutions to the single polynomial equation $x^4 - 2x^3 - 6x^2 + 2x + 1 = 0$.

Ramanujan's Tau Function

Ramanujan's innocently titled paper "On certain arithmetical functions" is arguably his most important work. The results in this paper include "congruences," which would later play a role in developing ideas that were critical to the proof of Fermat's last theorem, and it contained a conjecture that would later be proved by Deligne in his 1974 paper for which he was awarded the Fields Medal. Although it is too ambitious to try to explain these results in detail, it is worthwhile to see glimpses of what Ramanujan did.

Ramanujan defined a function that he denoted by the Greek letter tau (τ). It is obtained by multiplying infinitely many polynomials together. Now, that is exactly what is done on the right-hand side of the Rogers–Ramanujan identities shown above, where you can see an infinite product of polynomials in the denominator. Now, the only way for such a product to make sense is for the terms representing successively higher powers of the variable to stabilize. And that will happen only if any given power of the variable appears in only finitely many of the polynomials. Those conditions are satisfied by Ramanujan's function. That allows us to get more and more terms of the function by multiplying out more and more of the polynomials, as follows:

$$x\left(1-x\right)^{24} = x - 24x^2 + 276x^3 - \cdots + x^{25},$$

$$x\left(1-x\right)^{24}\left(1-x^2\right)^{24} = x - 24x^2 + 252x^3 - \cdots + x^{75},$$

$$x\left(1-x\right)^{24}\left(1-x^2\right)^{24}\left(1-x^3\right)^{24} = x - 24x^2 + 252x^3 - \cdots + x^{145}.$$

As one multiplies more of these polynomials, the terms at the beginning stabilize. Note that the second and third polynomials share the first three terms

$x - 24x^2 + 252x^3$, and every new polynomial from then on will begin with these three terms. If we were to multiply the next polynomial, then all the polynomials from then on would share the first four terms, and so on. In this way, Ramanujan was able to define a single object, a "power series" formed by multiplying infinitely many polynomials together. It is his "Delta" function

$$\Delta(x) = x(1-x)^{24}(1-x^2)^{24}(1-x^3)^{24} \cdots$$
$$= x - 24x^2 + 252x^3 - 1472x^4 + 4830x^5 - 6048x^6 - 16744x^7 + \cdots.$$

The tau function is now defined on the positive integers by taking the coefficients of the Delta function in order: $\tau(1) = 1$, $\tau(2) = -24$, $\tau(3) = 252$, $\tau(4) = -1472$, and so on.

Ramanujan proved that if p is prime, then $-1 - p^{11} + \tau(p)$ is without exception a multiple of 691. Here we show this for the first few small primes:

$$-1 - 2^{11} + \tau(2) = -2073 = -3 \cdot 691,$$
$$-1 - 3^{11} + \tau(3) = -176896 = -2^8 \cdot 691,$$
$$-1 - 5^{11} + \tau(5) = -48823296 = -2^{10} \cdot 3 \cdot 23 \cdot 691.$$

As strange as this may seem, it turns out that this phenomenon is a prototype of one of the deepest theories to be developed in the second half of the twentieth century, the theory of Galois representations, a universe that was imagined by Évariste Galois. How this leads to the proof of Fermat's last theorem is one of the longest and most beautiful adventures in the history of mathematics.

Ramanujan conjectured that his tau function is multiplicative, meaning that if m and n share no common prime factors, then $\tau(mn) = \tau(m)\tau(n)$. For example, if $m = 2$ and $n = 3$, then using the numbers above, we find that $\tau(2)\tau(3) = -24 \times 252 = -6048 = \tau(6)$. Ramanujan didn't prove this conjecture; it would later be proved by Louis Mordell, and it would go on to serve as the prototype of a central feature of modular forms, among the most important functions currently studied today.

Finally, Ramanujan formulated a conjecture about the rate of growth of the tau function. For every prime number p, he conjectured that $-2p^{11/2} < \tau(p) < 2p^{11/2}$. He thus conjectured that each of these numbers is restricted to a certain range. It shouldn't be too large or too small, and the conjecture makes precise what is meant by "large" and "small." This conjecture was further developed and folded into deep conjectures in algebraic geometry formulated by Weil, which were then proved by Deligne, earning him a Fields Medal.

Ramanujan invented his function tau in his 1916 paper, and the results he obtained seemed to be nothing more than oddities: strange-looking formulas and conjectured inequalities. Before him, nobody would have cared about this mathematics. It was his genius that recognized the value of these ideas, and it was up to mathematicians of the future to recognize their importance and make use of them. And so they did, and from the seeds that Ramanujan planted, a magnificent garden has grown.

Ken Ono and Dev Patel rehearsing the "Circle Method Scene" (photo by Sam Pressman)

Printed in the United States of America